Art in Motion III • Performing Under Pressure

Art in Motion III

Performing Under Pressure

ADINA MORNELL (Ed.)

PL ACADEMIC RESEARCH

Bibliographic Information published by the Deutsche Nationalbibliothek
The Deutsche Nationalbibliothek lists this publication in the Deutsche Nationalbibliografie; detailed bibliographic data is available on the internet at http://dnb.d-nb.de.
Library of Congress Cataloging-in-Publication Data
Names: Mornell, Adina, editor.
Title: Art in motion III : performing under pressure / Adina Mornell (ed.). Other titles: Art in motion 3 | Art in motion three
Description: 1st edition. | New York : Peter Lang, 2016.
Identifiers: LCCN 2016002984 | ISBN 9783631671726
Subjects: LCSH: Music--Performance--Psychological aspects. | Music--Performance--Physiological aspects. | Performance anxiety.
Classification: LCC ML3830 .A773 2016 | DDC 781.4/3--dc23 LC record available at http://lccn.loc.gov/2016002984

Editor:	Adina Mornell
Cover Art:	capdance (*Energy*, oil on canvas)
Reviewer:	Kelly O'Kane & Patrick Shal
Book Design:	capdance (www.capdance.com) *'For the Eye to Dance is Much Delight'*
Symposium Sponsors:	DFG Deutsche Forschungsgemeinschaft Hochschule für Musik und Theater München

ISBN 978-3-631-67172-6 (Print)
E-ISBN 978-3-653-06566-4 (E-Book)
DOI 10.3726/978-3-653-06566-4

PL Academic Research is an Imprint of Peter Lang GmbH.

Peter Lang – Frankfurt am Main · Bern · Bruxelles · New York · Oxford · Warszawa · Wien

This publication has been peer reviewed.

www.peterlang.com.

Table of Contents

Foreword

In the moment when I stepped onto the stage to greet the 400 attendees of the interdisciplinary and innovative conference "Art in Motion 2014" even I felt the pressure of performing in our large concert hall. Since English is not my native language, I tried to ease my stress by reading from a manuscript I had prepared. Nevertheless my heart was beating fast, and spontaneous thoughts disturbed my concentration. I didn´t sleep well the night before so I felt somewhat tired, though I was fully aware of the audience's high expectations. I wanted to appear competent and convincing, and was worried that if I failed, it might ruin my career.

All young musicians, dancers, actors and athletes face these problems in even more challenging situations. To cope with them is essential: An incredibly high level of perfection is demanded from each of them. The job market for performing artists is getting tighter and tighter. Competitive pressure is fierce. Many talented young artists and athletes are reaching the limits of the physiological and psychological load they can carry. This is the reason that medical prophylaxes are needed, based on the specific requirements of stage performers and athletes, as well as scientific, research-based learning and teaching strategies. Training for performance, presentation, and competition continues to become more and more important.

Institutions offering professional studies of music, dance, and theater are challenged to provide an intensive and direct network made up of research scientists, artists who teach, and teachers of the arts. This connectivity should be reflected in future curricula and structure of the university as well. Relationships and interdependencies have to be developed between the subjectivity of artistic choice on one hand, and the objectivity of a methodical, external and analytical perspective, on the other. An inclusion of both of these approaches will create an inspirational academic environment for artistic research.

Performing art without reflection and critical introspection is now obsolete. No young, unknown artist can afford to concentrate purely on creative work without considering the outside world. Both the highly sought after, but extremely rare permanent positions in cultural institutions (orchestra, opera house, theater), and jobs within public-supported educational institutions (e.g., music schools) offer less autonomy than being self-employed. On the other hand, these jobs provide a secure framework for artistic work, which eliminates the necessity for self-introspection. However, even musicians in orchestras and employees of the theater are increasingly expected to further educate and inform themselves, as well as to take responsibility for their own professional competence.

Thus, in preparation for a demanding labor market, today's students of music, dance, and theater need to develop complex skills within and beyond their main subject. Instrumentalists and singers, for example, cannot be content performing by memory with flawless technical perfection. In order to arrive at a high profile, originally distinctive interpretation, these artists must engage themselves intensively with the study of historical performance practice and the origins of the work, style, and genre of the piece. This requires an analysis of the structure of the work, an intellectual confrontation with the value of its artistic substance, and an understanding of the psychology of aural and visual perception. In order to find a niche for oneself as an independent artist, one must be equipped in the skills of self-management, know about the market, the repertoire, legal issues, entrepreneurial strategies, and have an understanding of the current methods of artist-audience communication, event organization, and the use of media. All of these are beyond the basic necessity that one speaks and writes about music. The self-employed need to be creative, produce innovative ideas, and be flexible and versatile, in order to develop their own products and methods of expression. They must keep up-to-date on the "scene" and interact with it, as well as be capable of adding a critical-reflective dimension to their creativity in order to develop artistic singularity.

In all of these areas one sees relationships and interdependencies between the subjectivity of artistic choice on one hand, and the objectivity of a methodical, external and analytical perspective, on the other. An inclusion of both approaches will create an inspirational academic environment for artistic research.

Based on these criteria for success in the performing arts, the University of Music and Performing Arts Munich hosted the interdisciplinary and innovative conference "Art in Motion 2014" in Munich. The symposium's theme "Performing under Pressure" reflects a field of greatest importance for every musician, stage performer, and athlete. While this book provides highlights of the symposium, it also includes additional chapters from performing under pressure experts whose research and opinions deserve to be part of this volume. The symposium set forth the series of biannual *Art in Motion* symposia that Adina Mornell brought to our university in 2010, and this book is the third in the series. Dear Adina, thank you for this great idea and fantastic work. I would also like to thank our partners for the collaboration that went into the symposium – the *German Association of Sport Psychology* (ASP) and the Department of Health Sciences of the *Technische Universität München* (TUM) – as well as the faculty, staff and students of our own university who contributed to its success. Let us hope that long-lasting interdisciplinary networks and joint research projects grow out of this symposium and this book. That would be an optimal way in which science could contribute to art.

Bernd Redmann
President
University of Music and Performing Arts Munich

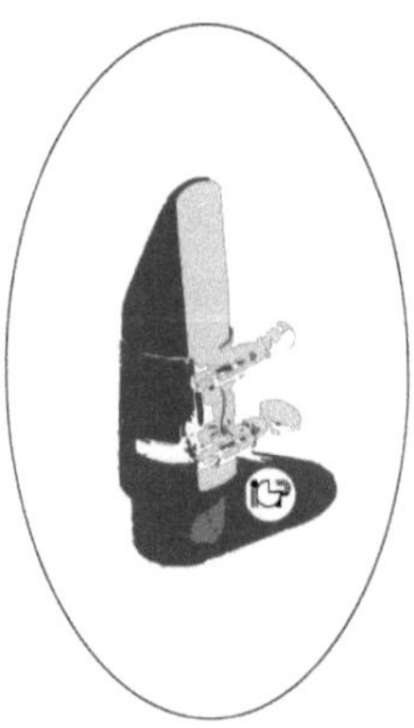

Internal Pressure During Musical Performances and Their Preparation

Personal Reflections With Drawings

"Actually, I was really well prepared for this concert and was looking forward to it. Then I climbed onto the podium. I began: beautifully, bravely, sonorously – all was well – until I suddenly and somehow got off track and lost my nerve. My ice-cold fingers managed to lose their bearings somewhere and somehow along the way. I was playing ever faster and faster, couldn't hold on anymore, lost my secure stance. It was as if the planks of the stage – those that mean the world to me – were suddenly giving way... the more I wanted to control, the stronger I was sucked into the vacuum of inconceivable energy that was coming from the beloved piece of music I was playing. What on earth did I do wrong?"

I hear such sentences rather frequently when teaching, and from a most heterogeneous group of students. Especially in Master Classes, where we are almost strangers, I still en-

Über den inneren Druck bei musikalischen Aufführungen und deren Vorbereitung

Persönliche Betrachtungen und Zeichnungen

„Eigentlich habe ich mich sehr gut vorbereitet und auf das Konzert gefreut. Dann kam ich auf das Podium, es war ein schöner, mutiger und klangvoller Start – bis ich plötzlich irgendwie den Faden verlor und meine Nerven auch. Meine eiskalten Finger haben sich in irgendwelche Passagen verirrt, ich wurde immer schneller und schneller, fand keinen Halt mehr, verlor auch meinen stabilen Stand, so als würden sich die Bretter, die mir eine ganz Welt bedeuten, plötzlich neigen... je mehr ich mich kontrollieren wollte, desto schlimmer saugte mich eine bisher unvorstellbare Energie aus der Musik heraus, die ich doch so liebe. Was mache ich nur falsch...?"

Nicht selten höre ich solche Sätze während meiner Arbeit mit den unterschiedlichsten Studierenden, gerade auch auf Meisterkursen, wenn man sich eigentlich noch kaum kennt und dennoch das Geschenk enormen

joy the gift of complete trust from the participants. Emotional stress, accompanied by an undertone of self-flagellation: this is what is evident in these very honest and open conversations. Upon closer examination, it appears to me that there's a repetitive pattern here: these complaints are about a perceived loss of control. It's interesting that most of the students arrive at this important point all by themselves. It's simply a question of what perspective they use to approach the issue. I dare say that hidden behind the catch phrase "loss of control" is the very opposite! It is all too often an "overdose" of control that's been acquired through practice, and it is this intentional and well-meaning "self-control" that exerts itself in stress situations – practice-acquired control is controlling us rather than us being in control of the situation. What this means for teachers like myself is that if I use sharp-tongued reprimands when I hear my students play this or that passage "for the umpteenth time with poor intonation," I would be responsible for establishing a deficit-awareness that – depending upon the personality of the student – will evolve into a bad mood, self-accusation, an unbalanced posture, or possible actual physical pressure. If, despite all of this, the student's intonation actually improves, this progress is nothing but a short-lived success, based upon

Vertrauens genießt. Evident ist der Leidensdruck, aber auch der Unterton der Selbstanklage während solcher offenen Gespräche. Bei näherer Betrachtung fällt mir dabei fast gleichlautend Folgendes auf: Beklagt wird vor allem der als solcher empfundene Kontrollverlust. Interessant, dass die Studierenden meist von selbst auf diesen kardinalen Punkt kommen! Es ist lediglich die Frage, mit welchem Blickwinkel diese Thematik angesteuert wird. Ich wage zu behaupten, dass sich hinter dem Schlagwort „Kontrollverlust" eher das Gegenteil verbirgt: Es ist allzu oft ein „Zuviel" der Kontrolle antrainiert worden, und diese vermeintliche, gut gemeinte „Selbstbeherrschung" beherrscht in Stresssituationen dann eher uns als wir sie. Konkret kann das bereits im Unterricht bedeuten: würde ich Studierende scharfzüngig zurechtweisen, dass etwas diese oder jene Stelle „mal wieder nicht sauber intoniert worden ist", würde ich in meinem Gegenüber ein Problembewusstsein etablieren, das sich je nach Charakterzug zunächst in schlechte Laune, Selbstbezichtigung, nicht ausbalancierte Körperhaltung bis hin zu physischem Druck übersetzt. Gelingt „dennoch" eine Verbesserung der Intonation, so ist sie eher dem kurzfristigen Erfolg einer militärischen Durchhalteübung gleichzusetzen. Doch das Opfer ist hoch: der saubere Ton wird „erkämpft", die innere

something that is indistinguishable from military drill. The sacrifice one makes to win the "fight" for the correct tone is enormous, as one's inner tension becomes part of the practice ritual.

The dilemma is, that this working method hides behind the guise of "indispensability": success is supposedly not possible without this form of misguided hard "work," consisting of endless repetitions of the same "trouble spots." This is typically an almost ritualistic behavior pattern: we start playing with a lively flourish in an unreduced tempo, and arrive at the "trouble spot," where we suffer a shipwreck and get angry. The operation is aborted. And already the "tree of trouble" has grown branches. First, the promise to play *through* the piece from start to finish is broken. Second, at full speed, we are not "in control" of this or that passage. Third, we swore that with this next attempt, all possible concentrated strength would surmount all obstacles. If physical effort is required in this passage, then a fourth – tension – follows hot on the heels of the previous three. Should there nevertheless be a noted short-term success, a dazzling self-deception (fifth) has been achieved, and this superficially soothes (sixth) our "conscience of work-ethic." Under these circumstances, how are we supposed to steady ourselves *for*

Verkrampfung bereits zuhause eingeübt.

Das Dilemma nun ist, dass sich diese Arbeitsweise als „unabdingbar" maskiert: als fehlgeleiteter „Fleiß", der aus endlosen Wiederholungen immer derselben „Problemstellen" besteht, ohne den angeblich kein Erfolg zu erringen wäre. Typisch hierfür ist ein geradezu rituelles Verhaltensmuster: Man nimmt erst einmal einen schwungvollen Anlauf im nicht reduzierten Tempo, erreicht die „Problemstelle", scheitert an ihr und ärgert sich. Der Vorgang wird abgebrochen. Unser „Problemstammbaum" hat sich schon jetzt bereits verzweigt: erstens haben wir unser Versprechen uns selbst gegenüber, das Stück mal wirklich *durch*zuspielen, gebrochen. Zweitens haben wir im vollen Tempo uns selbst gegenüber den Beweis erbracht, dass wir diese und jene Stelle nicht „beherrschen". Drittens schwören wir uns, beim nächsten Anlauf wirklich alle geballten Kräfte einzusetzen, um dennoch über die Hürden zu kommen. Kommt physischer Kraftaufwand ins Spiel, folgt viertens Verkrampfung quasi auf dem Fuße. Sollte dennoch kurzfristiger Erfolg zu verzeichnen sein, haben wir fünftens einen blendenden Selbstbetrug geleistet und sechstens unser „Arbeitsgewissen" oberflächlich beruhigt... Wie sollte man denn unter solchen Voraussetzungen den „Halt" *erstmals* auf

the first time ever on the stage? And how can we expect to achieve complete focus, when we have consistently lacerated our concentration, "thanks" to constant and arbitrary interruptions during the previous practice sessions?

If we oversimplify things and imagine our brain to be a storage unit that's hungry for everything we're ready to feed into it, we should be particularly careful about which "foods" we present it with, and which foods we do not. If, in the beginning, the "nuggets" are too big, we'll have to "break them down" before we can swallow them. And just as it is with actual eating, those who eat slowly, enjoy more, and digest better! We have to monitor what's eaten, and perhaps even *how* it's eaten!

At this point, I would like to, preemptively, address a possible misunderstanding. The previous example, taken from a typical music lesson on the theme of intonation, is not meant to rebuke teachers. Instead, it is intended to demonstrate, by using one exemplary detail out of thousands, that we walk an incredibly precarious fine line. Naturally, a precise examination of something like into-

dem Podium erwarten und eine Konzentration herbeiwünschen, die man sich selbst zuvor „dank" permanenten und willkürlichen Unterbrechens unsäglich zerschnitten hat?

Wenn wir uns stark vereinfachend unser Hirn als Datenspeicher vorstellen, der hungrig sei auf das, was wir ihm einspeisen, sollten wir sehr genau darauf achten, welche „Speise" wir ihm vorsetzen sollten und welche lieber nicht. Sind die „Brocken" zunächst zu groß, werden wir sie eben „zerkleinern", bis wir sie aufnehmen können. Und wie es auch beim Essen gilt: wer langsam isst, genießt *und* verdaut besser! Und beobachtet genauer, was gegessen, vielleicht sogar, *wie* gegessen wurde!

An dieser Stelle möchte ich einem drohenden Missverständnis vorbeugend begegnen: das voran stehende Beispiel aus herkömmlichem Unterricht zum Thema Intonation soll nicht Lehrende zurechweisen, sondern anhand eines von tausenden Details aufzeigen, wie enorm sensible der Grat ist, auf dem wir uns hier bewegen. Selbstverständlich wird eine präzise Untersuchung etwa der Intonation immer einen ganz wesent-

nation will be accorded great importance for string players, wind players, and singers. However, it is the "how" that is of importance here: will teacher and student be successful in hearing with a "mutual ear," scrutinizing how something sounds when projected into a room, and deciding which practice method can be adjusted to create the desired effect ... and then, without explicitly expressing it, successfully build a relationship of trust? Trust, not only in the teacher, but also – most of all! – trust in one's own actions. This is when mistakes stop counting as "enemies," and instead are seen as intense challenges that can be met without tension. In addition, this mindset allows one to create an "alter ego," whose bird's eye view strengthens one's ability to observe oneself. An amused attitude could be one wondrous product of this process. This amusement would remove unnecessary austerity from our work and help us see ourselves in the mirror. It's the work of self-reflection that summons the "inner teacher" in each of us. It is the occasional pause, a complete silence from time to time, that allows us to take a look at ourselves and enables us to discover our own highly personal and unique solutions.

Here is where the "public test" starts. Can I possibly experience that which I could at home in peace,

lichen Raum einnehmen bei Streichern, Bläsern und Sängern. Dabei ist das „*Wie*" entscheidend: Gelingt es, quasi „mit gemeinsamem Ohr" zu hören, gemeinsam zu hinterfragen, wie dieses oder jenes wirkt, das man tut, in den Raum projiziert, welche Übemethode typologisch angepasst diesen oder jenen Effekt hat, dann gelingt auch, ohne es explizit zu formulieren, ein wachsendes Vertrauensverhältnis: nicht nur zu den Lehrenden, sondern – vor allem! – zur eigenen Tätigkeit. Dann werden Fehler nicht zu „Feinden", sondern zu spannenden Herausforderungen, die es *ent*spannt anzunehmen gilt. In diesem mentalen Stadium kann sich zudem eine Art „alter ego" aufbauen, mit dessen distanzierter Optik der eigene Beobachtungsstatus entscheidend gestärkt wird. Eine fast amüsierte Attitüde kann die wunderbare Folge werden, die einer ernsthaften Arbeit die unnötige Strenge nimmt und sie uns spiegeln hilft. Es ist das selbst-reflektierende Arbeiten, das den „inneren Pädagogen" in uns wachruft, es ist das gelegentliche Innehalten, die völlige Stille zwischendurch, die uns ermöglicht, in uns zu gehen, um höchst individuelle Lösungswege zu erspüren.

Und hier setzt dann auch der „öffentliche Test" an: Kann ich mir das, was ich in Ruhe zuhause zugelassen habe, auch auf dem Podium ermöglichen? Wie genau erlebe

on the stage? How exactly do I experience the stage, and what happens to my breathing when I perform? How do I sense my feet; will I "endure" silence, or even welcome it, even as a moment of communal concentration before the first note sounds? Will I still experience my excitement as a moment colored by panic, something I believed up until now that I should fight against? Or is this excitement on the contrary a means of transportation for my momentarily elevated level of energy, that I would like to release, yes, even give away? I would preferably compare "practice" in preparation for performance with the process of packing for a long trip, in peace and with precision. All of the utensils are carefully selected and provide a sense of security for the future. However, in contrast to the travel analogy, I would rather leave my perfectly packed suitcase at home and find myself on the podium again – completely devoid of baggage! "Baggagelessness" symbolizes the freedom I've attained from the previously *and* lovingly acquired control, and for my focus on radical openness, spontaneity, and emotional flexibility. Those who have experienced what it means to break free from their own "high-security prison," especially in public, can sing a song of the endless horizon and freedom that they have enabled themselves to experience. Interestingly enough, mo-

ich das Podium, was macht meine Atmung beim Auftreten, wie spüre ich meine Füße, kann ich Stille vor den versammelten Menschen lediglich „aushalten“ oder begrüße ich sie sogar als geradezu gemeinsame Konzentration, noch bevor der erste Ton klingt? Erlebe ich meine Aufregung noch immer als panisch eingefärbtes Moment, gegen das ich bisher glaubte, ankämpfen zu müssen, oder ist sie nicht viel eher Transportmittel meiner momentan gesteigerten Energie, die ich nun endlich loswerden, ja: verschenken darf? Am liebsten würde ich das vorbereitende „Üben“ vergleichen mit dem Vorgang, in Ruhe und Genauigkeit meinen Koffer für eine lange Reise zu packen. Alle Utensilien werden sorgsam bedacht und assoziieren eine sichere Zukunft. Im Unterschied zur Reisethematik aber lasse ich nun meinen perfekt gepackten Koffer zuhause stehen und finde mich auf dem Podium wieder – völlig ohne Koffer! Es sei Sinnbild für die Befreiung von der zuvor penibel *und* liebevoll erfolgten Kontrolle und für die Ausrichtung hin zu radikaler Offenheit, Spontanität und emotionaler Durchlässigkeit. Wer immer es erlebt hat, was es bedeutet, ausgerechnet in der Öffentlichkeit seinen inneren „Hochsicherheitstrakt“ aufzugeben, weiß ein Lied zu singen von der unendlichen Weite und Freiheit, die zu schenken man dadurch imstande

ments of partial shortcomings are reduced in importance and do not drill any more holes into the once emaciated or however otherwise encrypted self-confidence. In this way, the commonly evoked "performance pressure" loses its damaging and constricting impact, and instead, can be transformed into an agent of change, propelling us beyond the expected, towards an highly successful performance, during which we performers keep our center, or start exploring and finding our center.

"I seemed to have played better than in any prior rehearsal – strange, because this time I didn't invest *as much* into practice the way I usually do ..."

These are the most beautiful, yet frequent, comments that I receive and take to heart. Naturally, they don't just make me happy! Those who speak this way have become a bit more conscious of themselves than they may have been previously. And they continue to work along this new approach with the sense of relaxation that goes along with it. Last, but not least, I do not want

ist. Interessanterweise reduzieren sich dann sogar Erlebnisse partieller Unzulänglichkeiten auf sich selbst und bohren keine generellen Löcher mehr ins einst ausgemergelte oder wie auch immer verschlüsselte Selbstbewußtsein. Dem vielbeschworenen Druck auf dem Podium kann auf diese Weise seine beschädigende und einengende Wirkung genommen werden, mehr noch: er kann zum Triebmittel einer kaum geahnten Gesamtleistung mutieren, in welcher wir als Ausführende unser Zentrum behalten oder gar erst richtig erkunden und finden.

„Ich habe eigentlich besser gespielt als in jeder Probe zuvor – komisch, denn dieses Mal habe ich doch gar nicht *so* wahnsinnig viel zuvor investiert an Üben wie sonst..."

Das sind dann die schönsten und regelmäßig geäußerten Sätze, die sich mir einprägen und natürlich nicht nur mich glücklich machen! Wer so spricht, ist sich über sich selbst ein wenig mehr bewußt als vielleicht früher und kann mit diesem neu gewonnenen Ansatz entsprechend entspannt weiterarbeiten. Ich möchte zu guter Letzt aber

to forget to mention something very important to me. When one takes big steps forward and experiences or undergoes transformation, a form of deep gratitude occurs that accompanies this kind of inner satisfaction: gratitude for all of this energy from which peace of mind draws its strength.

nicht versäumen, zu sagen, was mir sehr wichtig ist, wenn man gemeinsam enorme Schritte und Verwandlungen erlebt und durchlebt hat: Es ist eine Art tiefster Dankbarkeit, die sich einstellt und einhergeht mit Zu-Friedenheit. Dankbar für Energien, aus denen Frieden genährt wird.

Pedagogical Footnote:

When one only can, when one should, but never can, when one wants, then one will not want, when one must.

When, however, one can when one wants, then one will also want, when one should, and then also can when one must.

Those who should be able to, must also be allowed to want to.

(Source unknown)

Pädagogische Fußnote:

Wenn einer nur darf wenn er soll, aber nie kann, wenn er will, dann mag e auch nicht, wenn er muß.

Wenn er aber darf wenn er will, dann mag er auch, wenn er soll und dann kann er auch wenn er muß.

Diejenigen die können sollen, müssen wollen dürfen.

(Quelle unbekannt)

[Translation - with poetic license - by Adina Mornell]

Some of My Experiences With "Stage-Fright"

I was invited to participate in the very first "Art in Motion" Symposium, held in Graz, Austria (2008), and in the second symposium of the series in Munich (2010), and participated as an author in both of the books that grew out of those meetings. Now I have been invited to write an introduction for the third book. This invitation came from the initiator and director of those symposia, Adina Mornell, who somehow believed I had been an Olympic gymnast. (The truth is, I was *not* an Olympic gymnast at all, but only a reasonably proficient collegiate gymnast at UC Berkeley, specializing in the rings.) Nonetheless, I have agreed to participate with the following thoughts.

The motor-control literature contains several relevant discussions about factors that influence so-called "stage-fright" on the actions in skilled behavior necessary (e.g., for a high-level violin performance, including accurate fingering and bowing). As I have said previously, a considerable overlap exists between laboratory research in motor learning, motor behavior/control and high level music, voice, dance, and theater performances. This volume, *Performing Under Pressure*, is directed specifically toward bringing us up-to-date on how, or whether or not, this connection is also present for high-level music performances with regard to so-called "stage-fright." In asking me to write this preface, however, I've been asked to focus on instances of "stage-fright" in my own experiences.

When I was in fifth grade, a violin instructor (of some considerable repute), came to our fifth grade class to give a demonstration and to "plug" her new, local violin classes. I had *no* experience in music. Several of us in the class were intrigued, and, eventually three of us (two non-males and myself) signed up. The classes were held in the afternoon after school, were private, and were held locally. This stage concentrated on reading sheet music and bowing and fingering technic. We focused on fingering in the first position, with occasional visits even to the third position. At about this time, I inherited a very old violin that used to be my grandfather's.

A part of these classes was a terrifying "recital" before all the parents and guests whom they had brought along. My mom and dad (who had in his day become a very proficient band-trombonist) came as well. This was my first solo performance, and I remember that I was "petrified." Right after the recital, my dad managed to be positive, saying essentially, "You must have a decent 'ear' for music. Every time you made a fingering error, you detected and corrected that particular note on the fly while continuing to play." I guess the reader can understand why I gave up the violin classes shortly thereafter.

In junior high school though, one semester (based on the skills acquired through the lessons) I became associated with the San Fernando Junior High School Junior Orchestra. I eventually worked my way up to third violin (three rows up from the conductor on the side closest to the audience). This process continued for a semester or so; then the conductor one day announced that all orchestra members soon would have a solo examination, in order to re-assign the positions in the orchestra. This involved playing a section of music that we had worked on all by ourselves, with (of course) all the orchestra members watching. This was my first encounter with "stage-fright." I must have performed reasonably well as my position in the orchestra did not change.

The third, and certainly the most terrifying event, occurred at the University of California, Berkeley, when I was on the gymnastic team. The gymnastics program was relatively new, and so the gymnastics coach (Harold Frey) was offered the opportunity to put on a demonstration of the UC Berkeley (Cal) Gymnastics team (for promotional purposes), during half-time at one of the Cal basketball games. Cal had a pretty good gymnastics squad and basketball team that year (1962), so Harmon Gymnasium was packed with fans.

The demonstration began with the introduction of the whole team and a few words from Coach Frey. The rings "team" warmed up under the bleacher, "scoping out" the audience as we did. Then, half-time occurred. We went out to the gym's still rings apparatus and awaited our turn. When my time came, I was incredibly nervous and jumpy. I had the feeling from past experiences doing solo exercises (with the team watching) to know which moves I would likely miss under these circumstances, and so I began my routine. I performed relatively well, and did a dismount to get off the apparatus. I was certainly glad that was over! From my experiences training with the team, I knew which skill that I was apt to miss. I was so wrong. I did not miss anything. I chock that up to my several previous years competing in front of audiences.

Richard A. Schmidt on rings

The final experience that I had, occurred at the 1962 NCAA Gymnastics National Championships in Albuquerque. In this meet I had a real chance to win in the rings competition. I had even perfected a new dismount (a double fly-away, where the competitor does two somersaults and hopes to land under control). During the warm-up I had asked Jerry Todd (a very good gymnast coach at that time from Pasadena City College) to "spot me" on the double fly-away dismount. I performed a few stunts near the end of my routine, and then went for the double. Wouldn't you know it; I landed straight on my face/chest area! Todd must have thought that I was capable of doing the dismount, and did not spot me at all. Bereft of confidence, I immediately went to the warm-up area in the practice gym and practiced my old dismount, which I then used in place of the double fly-away in the routine. I did not win the rings that meet, but I tied for first place in the finals. Today I could pinpoint the cause of this failure as being cognitive in nature. My extreme nervousness, plus falling on my face caused me to make a bad decision. By eliminating the fly-away, I had catered to my fears by diminishing the risk in my performance. I would have profited from a book such as this one, which explores the phenomena of stage fright and choking under pressure.

C. Mesagno, A. Mornell, A. L. Quinn

Choking Under Pressure in Sport and Music: Exploring the Benefits of Theory Transfer Across Domains

Introduction

The environments in which musicians, singers, dancers, and athletes perform have some obvious differences, but they are also similar in many ways. One of the main commonalities is that in all of these performance arenas, individuals must be able to perform under pressure. If the performer is unable to deal with the pressure, the athlete may experience "choking under pressure" (i.e., choking) or the performing artist may suffer from performance anxiety or "stage fright," all of which are counterproductive to the individual's well-being, standard of achievement, and career. Researchers have developed attention-based theoretical explanations for choking under pressure in athletes. There is current and growing evidence in the sport sciences that explains why choking occurs when anxiety increases. Studies on music performance anxiety, on the other hand, tend to focus on physiology, personality characteristics, task-mastery, and external stressors. Within music psychology and performance research, only limited consideration has been given to attentional factors on stage. In this chapter, we will provide a comprehensive review of sport performance anxiety models and an overview of their applications to musical performance to date and suggest that a transfer of domain-specific knowledge be used between the fields. We will examine the development of anxiety, how attentional processes are impacted by an increase in anxiety, and when this can be detrimental to performance, both in music and sport.

Choking Under Pressure in Sport and Music

Anxiety in music performance is widely accepted as an occupational stress common to the musical profession (Yoshie, Kanazawa, Kudo, Ohtsuki, & Nakawaza, 2011). Frederic Chopin, famed composer of the romantic era, reportedly remarked on the stresses of performing "The crowd intimidates me and I feel asphyxiated by its eager breath, paralyzed by its inquisitive stare, silenced by its alien faces" (Samson, 1996, p. 29). State-of-the-art editing with advanced technology is used today to remove errors in the finished product, creating a "note-perfect" recording. Musical audiences come to a concert accustomed to this illusion of perfection, and their expectations of a polished and flawless performance place additional pressure on live musicians to reproduce these exceptional recordings. For classical pianists, performance from memory has been the norm since the 1800's (Bitzan, 2010), causing incidences of increased anxiety, especially since performance failures are remembered (Mishra, 2010), while successful performances fade quickly from memory. Many musicians can reflect on their performance career and identify their first major slip-up, or memory lapse (Kenny, 2011). Illustrious songbird Barbra Streisand suffered from an infamous memory lapse in a performance of popular music in New York in 1967, and struggled to return to the stage for 27 years following the episode (Marchesani, 2002). Mistakes are proverbially human: yet apparently not for singers and musicians who are expected to deliver perfection in the moment it counts (Sternbach, 1993). Thus, it is no surprise that this pressure is a source of conflict and even illness among even the most experienced performers (e.g., Casals, 1970; Tommasini, 2000).

Clearly, consequences of Music Performance Anxiety (MPA) can be devastating to performance and to the careers of musicians. Most of the symptoms associated with MPA are seen as detrimental, however, there are also benefits to the increase of excitement that comes with being on stage that have not yet been the subject of much research conducted within the psychology of music. There is no widespread agreement within the field about when musical achievement can be expected, or why and when errors occur. Both ends of the spectrum (i.e., *performance excellence* and *performance failure*) have thus not been fully explored. Within sport psychology research, when an athlete excels under pressure and performs better than expected, the athlete has experienced a "clutch" performance (Otten, 2009), whereas when an athlete suffers a collapse in performance due to an anxiety increase, it is called "choking under pressure" (i.e., choking; Mesagno, Harvey, & Janelle, 2011).

Choking Definition Debate

A universally accepted definition of choking has been elusive to sport researchers, and has been a topic of much discussion. Recently, Mesagno and Hill (2013) initiated the debate in a special "Performance under Pressure" issue of the *International Journal of Sport Psychology*, which was based on Hill, Hanton, Fleming, and Matthews (2009) choking study who questioned whether any performance decrement, or a substantial decrease in performance, should be classified as "choking." Mesagno and Hill (2013) argued for greater clarity in the choking definition with more investigations dedicated to examining if there are cognitive processing differences between under-performance and choking.

Irrespective of this debate, many theorists agree that choking involves three requisite conditions: execution of a skilled performance, perceived pressure that leads to increases in the performer's anxiety, and resulting performance degradation (Baumeister, 1984; Mesagno et al., 2011; Wan & Huon, 2005). Musicians may instantly recognize these three conditions as relevant to the performance of music, yet there is a dearth of literature on choking in musicians. In this chapter, we argue that music researchers could use sport anxiety research to develop theoretical knowledge about music performance.

A Comparison of the Domains of Sport and Music

The environments in which musicians, singers, dancers, and athletes perform have obvious differences, but also include many similarities, which have been highlighted previously (e.g., Yoshie, Shigemasu, Kudo, & Ohtsuki, 2009). Consequently, we have adapted our discussion of the comparison of sport and music domains from Yoshie et al., but have also added other conceptual similarities and differences.

Differences Between Sport and Music

The most prominent distinction between sport and music is that throughout its inception, practice, production, and consumption, music is grounded in creativity. Music has the ability to convey and evoke strong emotional reactions in performers and audiences. In music, feelings and emotions can be communicated through artistic expression and passion for what is being performed, which is not the case in sport. The aim in sport is to be more physically dominating than your opponent in order to achieve a higher score than the other individual or team, regardless of how it has been achieved or how the "audience" feels about it (although passionate team supporters may disagree). By contrast, a violist who is louder than the other players in his section of the orchestra is not a winner: he is out of place and potentially out of a job.

Another distinction is the subjective or objective nature of judgments made about the quality of performance. That is, music performances are evaluated subjectively, with researchers (e.g., Elliott, 1995/1996) indicating that music judges may vary their scores based on simple visual cues (Behne, 1988) such as gender. The simple presence of a music stand (even without sheet music on it) near the performer can cause a judge to mark the performer more harshly (Williamson, 1999). This evaluative subjectivity is also heavily influenced by societal pressure and expectation, which have increased enormously, along with the size of audiences, throughout the 20th and 21st centuries because of the dissemination of performances through the media. The presence of an audience sitting silently and staring at performing musicians is actually a relatively recent development (Hamilton, 2008). Even more current are the multitude of music-delivering devices (e.g., mp3 players, iPods, iWatches, etc.) that allow music to be heard outside the concert hall, practically everywhere and at any time. Yet, despite objectively note-perfect recordings created by advanced editing techniques, music is still inherently judged subjectively, as musical expression is more than just correctness. Within sport, technologies have also advanced to insure more objective measurement (e.g., with "hawkeye" technology now detecting whether a tennis ball is in or out and eliminate subjective judgment). Sport outcomes primarily rely on objectivity in their evaluation, and are simpler to measure, oftentimes determined by points scored by teams or individuals competing.

Finally, musicians are "small muscle athletes" (Altenmüller, Ioannou, & Lee, 2015), and use fine motor movements, requiring control of small muscles to achieve the goal of the skill (Magill, 2011) to creatively utilize the instrument, express the music they interpret, and create the desired tone quality (Altenmüller et al., 2015). A musician plans movement based on an internal model of the effect to be produced, then executes the action, receiving aural feedback and visual cues (e.g., when playing with the score or watching for the placement of one's hands). While planning the next phrase, the musician is evaluating the sounds just produced in a continuous feedback loop. Musicians must therefore work within the highest limits of their sensory and fine motor capabilities. Within many sports, in contrast, gross motor movements, which require control of large musculature to achieve the goal of the skill (Magill, 2011), are generally used to achieve the outcome needed for goal attainment. Athletes' mental, motor and sensory patterns are less complex. There are other, more obvious environmental differences between the domains, however, we argue that the similarities between the sport and music domains outnumber the differences and add to our argument that sport based knowledge should and could be applied to music research.

Similarities Between Sport and Music

The commonalities between sport and music can be classified as either being motor learning and motor skills, psychological skills based, social, or a combination of these. From a motor learning and motor skills perspective, both musicians and athletes need extensive training/rehearsal to perfect the motor skills to achieve a proficient level. This usually requires at least 10,000 hours of deliberate, purposeful practice on all motor learning skills (Ericsson, Krampe, & Tesch-Romer, 1993). During this behavior identified as "deliberate practice," athletes and musicians progress through stages of motor learning (Fitts & Posner, 1967). Both transition from the cognitive stage, where the skill is learned through considerable cognitive activity to performing the task, toward the autonomous (procedural) stage, where skills are more implicit (i.e., the musician functions with less awareness and the athlete with minimal effortful cognitive processing; Ericsson, 1996). As learning occurs and automaticity increases, less attention is needed for explicit rules of functional movement, leaving sufficient attention available to process additional relevant, but also irrelevant, information. Furthermore, athletes and musicians are similar in their use of memory. Whereby athletes may remember and recall different offensive and/ or defensive formations, musicians remember and recall the musical scores, hand positions, fingerings, and special techniques required to interpret a particular piece of music.

From the social element, sport and music are similar in that both necessitate good communication and teamwork but also involve individuals to do their part for the needs of the team/ ensemble. An environmental likeness between both music and sport that might lead to psychological similarities, including increases in psychological pressure, is that they are both played in front of an audience. As a result, psychological factors resulting from similar social evaluative situations may exist, which in both domains lead to increases in performance, somatic, and cognitive anxiety. Thus, individuals in both fields must be able to perform at their best under pressure and are equally vulnerable to social phobia. Since increases in anxiety occur during performance, it is therefore important for artists and athletes to train psychological skills such as mental imagery, concentration skills, or positive self-talk in order to excel in their domain.

This brief discussion about the differences and similarities between sport and music does not provide an exhaustive list, but rather a simple comparison to illustrate the overlap and demonstrate that sport anxiety research could be applied to music research. In summary: Musicians and athletes are performers in distinct, but similar environments.

Anxiety-Performance Relationship in Sport

Anxiety is often divided into two broad categories: state and trait anxiety (Jones & Swain, 1995; Martens, Vealey, & Burton, 1990). Spielberger (1966) developed the state-trait theory of anxiety, as an attempt to differentiate between anxiety as a mood state and a personality trait. *State anxiety* is an emotion characterized by subjective, consciously perceived feelings of apprehension and tension, accompanied by activation or arousal of the autonomic nervous system, which may vary from moment to moment and would fluctuate proportionally to the perceived threat in the immediate situation (Spielberger, 1966). In short, state anxiety is higher in conditions of perceived pressure (e.g., competitions, auditions) and lower in practice situations, where performance outcomes "don't matter" as much.

Spielberger (1966) also explains that *trait anxiety* is an acquired behavioral disposition that predisposes an individual to perceive a wide range of objectively non-dangerous circumstances as threatening. High trait anxious individuals perceive more situations as threatening and react with greater state anxiety than low trait anxious individuals. Sport anxiety researchers have confirmed that trait anxiety is a strong predictor of state anxiety (e.g., Marchant, Morris, & Andersen, 1998; Williams & Krane, 1992). Halvari and Gjesme (1995) also found that trait anxiety was related to both state anxiety and performance errors. That is, high trait anxiety was accompanied by associated increases in state anxiety, which was depicted by a strong U-shaped relationship to performance errors with moderate state anxiety associated with fewer errors than low or high state anxiety (Broadhurst, 1957). Thus, generally, high trait anxiety leads to increased state anxiety, which may impair performance.

Trait and state anxiety have thus been accepted concepts in psychology for many decades. In the following, additional forms of anxiety, related factors and potential causes of performance degradation will be introduced. We have included Figure 1 below for ease of understanding of the content presented in this chapter. This figure indicates relationships between the types of anxiety, plus the effects that anxiety has on attention, as these might impact choking susceptible athletes and, thus, potentially influence MPA susceptible musicians as well. There may be other personality characteristics related to choking that are beyond the scope of the discussion of this chapter (see: Mesagno, Geukes & Larkin, 2015 for a review). Therefore, the figure shows the trajectory from predispositions via signs of perceived pressure to conscious or unconscious cognitions that may create performance degradation.

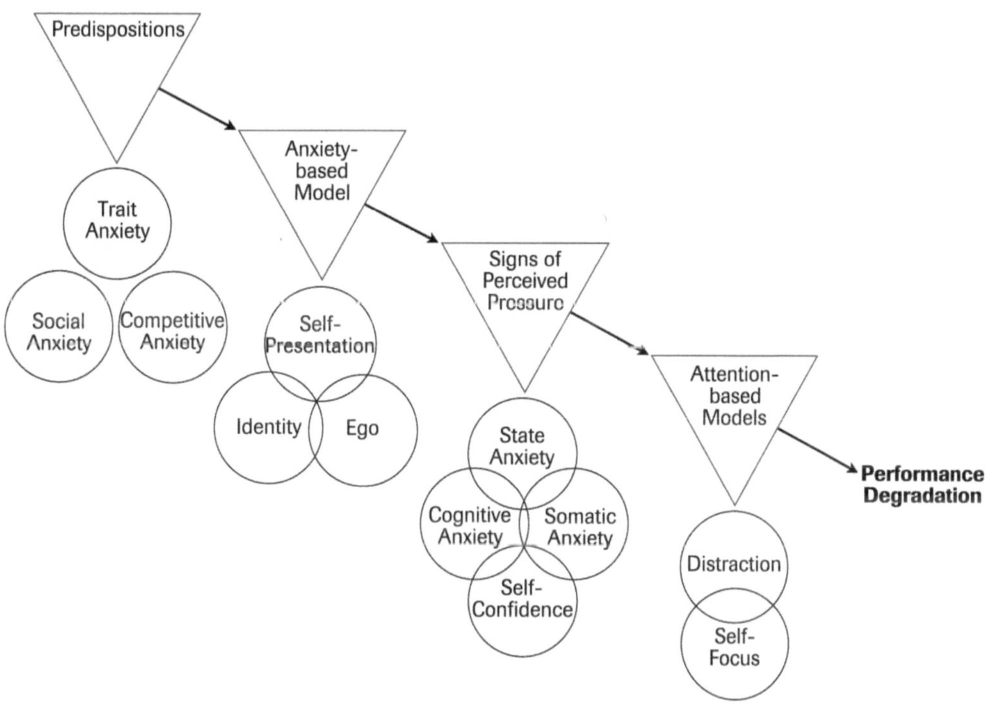

Figure 1: Factors Potentially Influencing Choking-Susceptible Performers

According to Martens et al. (1990), state anxiety is a multidimensional construct, consisting of two anxiety components primarily from a psychological (cognitive anxiety) and perceived physiological (somatic anxiety) perspective. Morris, Davis, and Hutchings (1981) first proposed original, operational definitions of cognitive and somatic anxiety. Morris et al. defined cognitive anxiety as "negative expectations and cognitive concerns about oneself, the situation at hand, and potential consequences" (p.

541). Cognitive anxiety is characterized by a state of worry, the awareness of unpleasant feelings and concerns about the ability to perform and concentrate in a particular environment. Whereas cognitive anxiety represents the psychological component associated with unpleasant cognitions, somatic anxiety is the perceived physiological component associated with multidimensional anxiety. Morris et al. defined somatic anxiety as "one's perception of the physiological-affective elements of the anxiety experience, that is, indications of autonomic arousal and unpleasant feeling states such as nervousness and tension" (Ibid.). Somatic anxiety is represented in perceptions of physiological processes such as shakiness, sweating, increased heart rate, rapid respiration, and "butterflies" in the stomach (Rotella & Lerner, 1993).

Multidimensional Anxiety Theory (MAT; Martens et al, 1990) is one of many anxiety-performance theories (e.g., Drive Theory, Inverted-U Hypothesis, Catastrophe Theory) that have been proposed in sport. We will focus on MAT because it may relate most to music performance anxiety and our discussion. MAT hypothesizes that there are three conceptually different components (i.e., cognitive anxiety, somatic anxiety, and self-confidence), and are related to performance in seemingly distinctive ways. This orthogonal anxiety theory predicts that cognitive anxiety and somatic anxiety differentially affect sport performance (Burton, 1988). More specifically, cognitive anxiety is posited to exhibit a negative linear relationship with performance; the more athletes worry, the greater the detriments to performance. Somatic anxiety is expected to have an inverted-U shaped relationship with performance (see M. Osborne in this volume, p. 93ff.). Performance is optimal at moderate levels of somatic anxiety, but performance decrements occur if moderate somatic anxiety is not reached or is surpassed. Self-confidence has a positive linear relationship with performance; self-confidence increases linearly as performance increases. The MAT suggests that athletes will perform at optimal levels when competing with high levels of self-confidence, moderate levels of somatic anxiety, and low levels of cognitive anxiety.

Competitive Anxiety and Performance in Sport

Sport is a social domain in which pressure is amplified possibly because athletes become aware of, and concerned with, factors such as playing in front of a large audience, possible de-selection on a sporting team, ridicule from the media, coaches, opposition, and spectators, and losing sponsorship endorsements. Researchers have indicated that athletes often experience intense competitive anxiety as a result of these extraordinary pressures accompanying elite sport (Mesagno et al., 2011). Competitive anxiety is defined as an individual's tendency to perceive competitive situations as threatening and respond with an increase in state (situational) anxiety (Martens et al., 1990), which may be a result of social evaluative processes and apprehension. Competitive anxiety may originate from numerous sources including social scrutiny from players, coaches, spectators and the media (Martens et al., 1990; Mesagno et al., 2011). Possible qualitative, social evaluation themes that usually emerge as sources of stress when competitive anxiety is present include (but are not limited to): "others' expectations," "pressure to skate up to national championship standards," and "pressure to be better than previous performance" (e.g., Gould, Jackson, & Finch, 1993; Martens et al., 1990; Scanlan, Stein, & Ravizza, 1991). Due to the perceived social evaluation that develops during competitions, Leary (1992) has conceptualized competitive anxiety as a sport-specific class of social anxiety.

Competitive anxiety can manifest itself as cognitive, somatic and behavioral symptoms, which can affect an athlete's functioning and ultimately, ability to perform to an optimum level (Bar-Haim, Lamy, Pergamin, Bakermans-Kranenburg, & Van IJzendoorn, 2007). Behavioral symptoms, such as rushing a routine or turning your back on the stressor/ goal-keeper in soccer penalty kicks, may be the result of heightened cognitive, competitive anxiety (e.g., Jordet & Hartman, 2008). Although the underlying mechanism of why is still unclear, researchers (Baumeister, 1984; Martens et al., 1990; Masters, 1992; Woodman & Hardy, 2001) generally agree that competitive anxiety contributes to decreases in performance under pressure. Competitive pressure has also been shown to affect other sport experiences including discontinuation of sporting activities (Scanlan, Babkes, & Scanlan, 2005), and less pleasurable experiences while participating in sports (Scanlan et al., 2005).

Anxiety-Performance Relationship in Music

Most musicians would agree that an increase in arousal is necessary for peak performance. Some languages reflect this in their respective terms for stage fright that indicate a slight rise in temperature, such as "fever under the light" in the German word *Lampenfieber* and the Spanish *fiebre de candilejas* (Spahn, 2012). The English term "stage fright" does not reflect this nuance of elevated excitement impartially, but rather gives a negative valence to the condition through the word "fright." The question that arises is: Does arousal (i.e., activation or excitement) turn into anxiety and fear at a particular level or under particular circumstances? In other words: are MPA and stage fright synonyms for the same condition, or is MPA an extreme form of stage fright?

Some argue that MPA is a condition in its own right (e.g., Möller, 1999). As summarized by Kenny (2011), Steptoe (2001) differentiates stage fright from MPA by listing three main points: first, stage fright can affect any performing artist (e.g., actors, dancers, and singers), whereas MPA affects only musicians; second, stage fright is generally a reaction to a large audience (Jackson & Latané, 1981); and third, stage fright occurs very suddenly and as a direct consequence of performing in front of an audience, whereas MPA can build weeks before a performance. In these ways, stage fright may be conceptualized as analogous to somatic anxiety, as a direct and sudden response to an external stressor, and as a subset of an MPA experience (Kenny, Davis, & Oates, 2004). Not all of these three points find agreement within the musical scientific community. Audience size as a stressor, for example, is a controversial issue and may not be an ideal determinant to discriminate between stage fright and MPA. Some researchers have provided evidence that audience size does *not* predict MPA (Brown & Garland 1971; LeBlanc, Jin, Obert, & Siivola, 1997), and at least one study shows some musicians even play *better* (Desberg & Marsh, 1988) in the face of a full auditorium. This is supported by a study with experienced musicians that described decreases in anxiety leading to a decrease in performance quality (Hamann, 1985).

Given that musicians may experience distress even in the very intimate setting of an instrument lesson (i.e., one-on-one) or a living room with just one or two friends (Havas, 1973), the fact that someone else is listening (and potentially evaluating) may lead to anxiety. Salmon (1990) offers a MPA definition that focuses on the musician's discomfort and the degradation of performance that occurs when anxiety is present, thus:

> MPA is defined as the experience of persisting, distressful apprehension about and/or actual impairment of performance skills in a public context, to a degree unwarranted given the individual's musical aptitude, training, and level of preparation (Salmon, 1990, p. 3).

Although this definition seems to encompass a large range of symptoms, it is clearly focused on the negative effects of arousal. Brodsky (1996) suggests the use of the diagnostic label *Music-Performer's-Stress-Syndrome (M-PSS)*, with the intention that a "syndrome" would allow room for the healthy aspects of stress and anxiety.

To date, the examination of MPA has been conducted using diverse methods based on differing definitions, which may lead to scientific and artistic confusion (Mornell, 2002). Some researchers, including Kenny (2011), view MPA as a combination of cognitive and somatic anxiety factors, and predominantly a form of state anxiety. Kenny (2010) defines MPA as a state-based anxious apprehension experienced within the context of musical performance, developed as a result of biological and psychological conditioning experiences, and that can affect musicians of all ages and levels of experience.

> It is manifested through combinations of affective, cognitive, somatic, and behavioral symptoms. It may occur in a range of performance settings, but is usually more severe in settings involving high ego investment, evaluative threat (audience), and fear of failure [...] It may or may not impair the quality of the musical performance (Kenny, 2010, p. 433).

Other researchers point to MPA as being influenced by a number of factors including personal (e.g., trait), situational, and environmental (Ayres, 1990; Brotons, 1994; Hamann, 1982); some experts also consider MPA to be a special type of social phobia (Cox & Kenardy 1993; Simoens, Puttonen, & Tervaniemi, 2015) or special form of social anxiety (Nicholson, Cody, & Beck, 2015). The characteristics of the musician may also influence their susceptibility to anxieties and negative outcomes, leading to an explanatory model for MPA that incorporates personality factors and trait anxiety (Dews & Williams, 1989; Papageorgi, Hallam, & Welch, 2007).

Almost all the proposed definitions for MPA draw on the same theoretical knowledge used in sport (state, trait, and cognitive anxiety) and offer no conclusive description or add no music-specific insight into the phenomenon (Brodsky, 1996; Mornell, 2002). In summary, efforts to reign in the experimental data, propose a unified definition, and organize the causes of MPA (including: Brandfonbrenner, 1999; Brugués, 2011; and Kenny, 2010) have served rather to extend the compendium of symptoms and contributory factors, rather than to refine it.

One thing that musicians and music psychologists can agree on is that, regardless of the term used, stage fright/ MPA is multi-dimensional and omnipresent in the life of a musician. It can occur across the lifespan, affecting children and adults (Boucher & Ryan, 2011) of both genders. In a review of the epidemiological literature, Brugués (2011) found that, in general, the onset of MPA appears to correlate with puberty, with adolescents demonstrating similar MPA symptoms to adult musicians. MPA has also been found to have a variety of predictors. For example, gender may be a moderating variable, with MPA being higher in females than in males (LeBlanc et al., 1997; Osborne & Kenny, 2008). Perhaps unsurprisingly, solo performances are much more anxiety provoking than ensemble performances (Cox & Kennardy, 1993; Jackson & Latané, 1981). Thus, MPA can affect many people of all ages and sexual orientation, especially when the performer is a soloist.

Musicians will endure a certain level of discomfort for their art, which does not keep them from performing, even over a lifetime of MPA. For example, cello maestro Pablo Casals was over the age of 80 when he published his memoirs (1970), admitting that he suffered from stage fright every time he performed. This is just one example that despite the advice commonly given to musicians to "just perform more often," an increase in experience does *not* result in a decrease in MPA (Cox & Kennardy, 1993; van Kemenade, Son & Heesch, 1995). Attempts to use systematic desensitization, a commonly recommended treatment for anxiety (e.g., Wine, 1982), to reduce MPA were not conclusively effective (Appel, 1976). Another counter-intuitive fact is that those who appear to be over-anxious may face the stage more successfully than their self-secure counterparts, having done the "work of worrying" (Burstein & Meichenbaum, 1979; Fenz & Epstein, 1967) as part of their pre-performance practice, which may essentially inoculate themselves from stress and prohibiting panic on stage (Epstein, 1983; Meichenbaum & Cameron, 1983).

Whether MPA is classified as chronic stage fright or not, directing attention to the phenomenon has been proven to be more beneficial than denial of the problem or avoidance of performing situations, as demonstrated by studies in which specific, additional practice strategies (musical analysis: Sweeney & Horan, 1982), various forms of coping (cognitive therapy: Kendrick, Craig, Lawson, & Davidson, 1982; behavioral therapy in the form of cognitive restructuring: Clark & Agras, 1991), and relaxation through guided imagery (Hammer, 1996), just to name a few examples, aided musicians' performance security and relieved perceived MPA.

Competitive Anxiety in Music

Competitive anxiety in music has yet to be systematically examined, although such a study would appear promising. Many of the themes described in sport sciences are identical to the domain of music. Parallel to the competitive anxiety experienced by athletes, musicians, too, perceive competitive situations such as auditions for an orchestra chair or international competitions as threatening. Naturally, state anxiety increases under these circumstances and can lead to a downward spiral of cognitive appraisal that includes fear of negative judgment by the jury (e.g., judging panel) or the audience, as well as fear of failure and loss of face.

Threatening situations can be found in the fierce competition for jobs and thus their livelihood among popular musicians (Cooper & Wills, 1989). Social scrutiny and fear of negative evaluation is inherent in public performance for both music and sports, while the fear of negative evaluation in advance of a concert (Kenny, Fortune, & Ackerman, 2011; Nicholson et al., 2015) can provoke as much anxiety as the actual event itself (Salmon, Schrodt, & Wright, 1989). A stressor identified in music that is likely to also be present in sports is the fear of not living up to the demands of one's country and/or culture (McPherson, 2009; Pomerantz, Grolnick & Price, 2005), and the burden of "great guilt" (Kempinska-Welbel, 1995) that musicians feel due to weight of their parents' expectations. Young musicians feel especially driven to work hard and succeed to repay the "sacrifices made by their parents, ostensibly on their behalf" (Chua & Rubenfeld, 2014, p. 109). Competition creates anxiety when one fails to meet one's own and others' expectations of performance success.

An athlete suffering from competitive anxiety who loses or does not enjoy competition because of the stress that accompanies it may quit participating in the sport. Musicians, however, continue to perform, often despite their level of discomfort in front of an audience, distress in performances, or ability to view concertizing as a positive experience. Competitive anxiety may not be as prevalent in musicians as in athletes, which may be due to the fact, viewed over their lifetimes, musicians participate in fewer competitive situations than athletes. Music competitions are one subset of a variety of performance venues open to musicians.

Another factor may be that musicians experience little direct conflict in competition, since in music contests each competitor is rated individually and then scores compared. Also, musicians compete early in their career to gain recognition, prize money, to obtain an orchestral position, so competitions usually become less and less a part of their life as they age and develop a name for themselves. In contrast, in sports, the more famous one is, the more exposure one has to competitive situations. Nevertheless, music psychologists may be able to learn from choking models within sport. We suggest that an exploration of "choking under pressure" will increase musicians' awareness of the spectrum of human behavior during performance, as well as add empirical knowledge to the discussion while providing a framework for future research.

Choking Models in Sport

Because important situations in sport typically produce increases in both somatic and cognitive anxiety (Baumeister, 1984; Martens et al., 1990; Marchant et al., 1998), accompanied by behavioral and attentional changes, some athletes are unable to perform up to previously exhibited standards and experience "choking." The causes and mechanisms of choking in sport have led researchers to examine attention-based models, but there is also emerging evidence for an anxiety-specific model of choking that might explain why athletes become anxious, which may be directly relevant to musicians. All categories of choking models may be applicable, but may be dependent on the type of performance, task-difficulty, and level of expertise (Hemery, 1986).

Anxiety-Based Choking Model in Sport: Self-Presentation

Considering that one essential component of choking is an increase in anxiety, it seems peculiar that sport anxiety researchers have focused more on attention-based models than on anxiety-based explanations of choking. Only recently have sport anxiety researchers (Mesagno, 2009; Mesagno et al., 2011) begun to study anxiety origins of choking. Mesagno (2009; Mesagno et al, 2011) proposed that self-presentation may be the genesis of concerns that increase anxiety for choking-susceptible athletes (i.e., athletes more likely to experience choking). Self-presentation is the process by which people attempt to monitor and control how they are perceived and evaluated by others (Schlenker, 1980). Mesagno's (Mesagno, 2009; Mesagno et al., 2011) self-presentation model is grounded in Leary's (1992) suggestion that competitive anxiety is a sport-specific class of social anxiety. Athletes that are most susceptible to self-presentation and choking want others to view them in a desirable way; therefore, they attempt to control others' judgments made about them by coordinating their cognitive processes and behaviors in ways that convey a positive image and help minimize social anxiety.

The self-presentation model was initially based on 14 interviews (Mesagno, 2009) with choking-susceptible athletes who participated in separate mixed-method research designs. Analysis of participants' interviews after the quantitative component indicated a link between self-presentation and choking, explained through public self-consciousness (i.e., the tendency to focus on outwardly observable aspects of the self) and fear of negative evaluation. Since then, quantitative support for the central role of self-presentation has been found (e.g., Geukes, Mesagno, Hanrahan, & Kellmann, 2013; Mesagno et al., 2011; Mesagno, Harvey, & Janelle, 2012). For example, Mesagno et al. (2011) found that pressure situations that involve public evaluation increased athletes' anxiety and decreased performance more than motivational pressure manipulations.

In a follow-up study, fear of negative evaluation was found to be inversely associated with performance (Mesagno et al., 2012). Fear of negative evaluation is defined as:

> apprehension about others' evaluations, distress over their negative evaluations, avoidance of evaluative situations, and the expectation that others would evaluate oneself negatively (Watson & Friend, 1969, p. 449).

Mesagno et al. (2012) conducted a basketball free-throw study in which athletes high (compared to those low) in fear of negative evaluation showed a significant increase in anxiety and a decrease in performance from a low-pressure to a high-pressure laboratory-based situation. Additionally, cognitive anxiety partially mediated the fear of negative evaluation and performance relationship, whereby extensive worries have an additional detrimental effect on performance under pressure (Mesagno et al., 2012). Explaining these results, Mesagno et al. (2012) suggested that the personality trait "athletic identity" might play an important role in determining athletes' self-presentation concerns. Brewer, van Raalte, and Linder (1993) defined athletic identity as the degree (i.e., strength and exclusivity) to which individuals identify with the role of an athlete.

Mesagno et al. reasoned that competitions (i.e., high-pressure situations) might create self-presentation worries (e.g., about making a good impression or "losing face") especially when involving social comparison or a threat to athletic identity. Thus, if athletic identity is a primary factor of athletes' self-concepts, performances in settings that clearly refer to the athletic role and are important to the athlete should increase anxiety.

Applying Leary's (2001) relational devaluation construct (i.e., people experience an increase in social anxiety when they believe impressions made will lead others to devalue or avoid relationships with them), if an individual's (sport or athletic) identity is under threat in a pressure situation, then an increase in anxiety may occur because of the possibility for relational (and sport-specific) devaluation. Furthermore, the more strongly and more exclusively individuals identify with the role of an athlete, the more they may depend on positive outcomes achieved within their athletic role to protect and maintain levels of self-esteem.

The self-presentation model is based on the premise that to be recognized as a successful athlete mandates the ability to perform well under pressure, but in sport, however, performance (observable sport behavior and outcome) is often the only method of conveying an image to the public. Choking-susceptible athletes do not usually enjoy being "in the spotlight" because they fear that their public image as an athlete will be devalued if they perform unsuccessfully. The choking-susceptible athlete is likely high in athletic identity exclusivity and is limited on self-complexity (i.e., the number of self-representations they possess; Linville, 1985), which is likely because of the tremendous amount of time and energy invested about being an elite athlete, limits their identification with other dimensions of themselves. This could influence the emotional reaction, including perhaps strengthening anxiety experiences, during the competition. An increase in self-presentation cognitions during performance may occur because self-, and others', positive appraisal may be diminished as a result of ego threat. Self-presentation cognitions may divert attention from task execution and lead to attentional shifts (via the self-focus or distraction models) during high-pressure situations in the hope that performance will be facilitated and also minimizing the likelihood of a devalued public image, which eventually leads to diminished performance. The model has been primarily discussed from an athlete and sport performance perspective; however, it is applicable to other choking scenarios that involve domain-specific identities (e.g., actors, dancers, musicians, singers, and surgeons), skillful movement outcomes, and the potential for increased social anxiety.

Attentional-Based Models in Sport

Within sport, performance excellence develops when the athlete voluntarily focuses attention on relevant information, processes, and behaviors, while concomitantly ignoring irrelevant and unwanted distractions (Nideffer, 1992). According to Nideffer, voluntary focus on task-relevant cues assists in maintaining optimal performance. Attention-based models of choking indicate that performance decrements occur when the athlete diverts attention to either internal (i.e., movement-related or emotion-focused) or external (i.e., environmental) irrelevant cues (Wulf, 2013) because of increases in anxiety (Nideffer, 1992). These unnecessary attentional shifts lead to sub-optimal concentration and performance decrements. These attention-based models can be classified as either distraction or self-focus models of choking.

Distraction Model

Distraction models originated within test anxiety research (Wine, 1971), where researchers found that poor test taking was a result of shifts in attention from task-relevant (e.g., the question being answered) to task-irrelevant (e.g., worry about outcome or the situational demands) information. Distraction model advocates (e.g., Eysenck & Calvo, 1992; Eysenck, Derekshan, Santos, & Calvo, 2007; Gucciardi, Longbottom, Jackson, & Dimmock, 2010; Oudejans, Kuijpers, Kooijman, & Bakker, 2011) suggest that choking occurs because attention shifts from task-relevant to irrelevant cues as a result of heightened anxiety. Choking occurs because athletes shift their attention from task and important sport cues to non-sport specific information.

Furthermore, Processing Efficiency Theory (Eysenck & Calvo, 1992) and Attentional Control Theory (Eysenck et al., 2007) are thought of as distraction models while not sport-specific choking models. Processing Efficiency Theory (PET) supporters suggest that worry leads to a reduction in the storage and processing capacity of the working memory (i.e., a short-term memory system that maintains, in an active state, a limited amount of information; Kane & Engle, 2000), as well as increases effort designed to improve performance (Eysenck & Calvo, 1992). PET supporters differentiate between performance effectiveness (i.e., quality of performance) and processing efficiency (i.e., performance effectiveness divided by effort) with a major difference being anxiety impairs efficiency more than effectiveness (Eysenck & Calvo, 1992). Thus, increases in anxiety result in decreases in processing efficiency because of extra effort and reduced attentional resources. Yet performance effectiveness could be maintained with extra effort, or impaired despite extra effort. Attention Control Theory (ACT) posits that anxiety increases as a result of a threatened goal, which causes attention to toward detecting the threat and deciding how to respond (Eysenck et al., 2007). Consequently, cognitive processing is likely to be involuntarily diverted toward task-irrelevant cues (Eysenck et al., 2007). ACT supporters postulate that the shifting and inhibition functions of the central executive of working memory are most adversely affected by anxiety. The inhibition function uses attention control to fight off task-irrelevant interference, whereas the shifting function allows shifts and maintains focus on task-relevant stimuli (Eysenck et al., 2007).

Self-Focus Model

Advocates of self-focus approaches (e.g., Beilock & Carr, 2001; Jackson, Ashford, & Norsworthy, 2006; Masters, 1992) have adopted Baumeister's (1984) Automatic Execution Hypothesis, which explains that choking occurs when the athlete allocates explicit information to movement execution as a direct result of increase in anxiety. Baumeister conducted a series of studies that demonstrated that increased self-focus was detrimental to performance and that pressure decreased performance in an experimental and field based study. Masters (1992) then extended Baumeister's work into explicit (i.e., formal rule-based instructions about how to perform the skill) and implicit (i.e., no explicit instructions provided) motor learning styles, where Masters suggested that the method in which participants learn motor skills may impair performance under pressure. Masters' Conscious Processing Hypothesis (or Reinvestment Theory) indicates that anxiety encourages attentional shifts toward conscious processing of explicit performance rules even if the experienced or expert performer does not need to refer back to earlier stages of motor learning. Implicit motor learners are more likely to perform automatically under pressure than explicit motor learners, thus, not relying on "reinvestment" in explicit rules.

Beilock and Carr (2001) also sought to document the accessibility of declarative (i.e., formal, rule-based [Anderson, 1987]) knowledge about golf putting with changes in skill learning. Beilock and Carr hypothesized and found that generic knowledge increases as level of skill progresses, whereas episodic knowledge decreases with further expertise. Skilled putting is encoded into a procedural form, which should allow performance to occur without the need for constant attentional control. Jackson et al. (2006) also extended these results by explaining that a breakdown in motor skills occurs when an athlete consciously monitors *and* controls movements, rather than monitors movements alone.

At a similar time, Wulf, Höß, and Prinz (1998) performed a classic motor learning and attentional focus study demonstrating that external focus of attention facilitates and internal focus of attention impedes performance. Both Wulf, McNevin, and Shea's (2001) Constrained Action Hypothesis (CAH) and Beilock and Carr's Explicit Monitoring Hypothesis (EMH) advocate that awareness of step-by-step procedures when executing well-learned behaviors is disruptive to high-level skills that have become automatized and function with little conscious effort (Duke, Cash, & Allen, 2011). The main differences for these hypotheses and investigations are that supporters of the CAH examine learning with the focus on attentional direction without the necessary prerequisite of anxiety. Furthermore, EMH are interested in the underlying mechanisms for why attentional shifts toward conscious processing when anxiety increases. Nevertheless, these studies substantiate the self-focus model of choking during skilled performance.

Music Performance Anxiety Models and Choking in Music

The lack of a universally accepted definition has not hindered empirical research on MPA, although this research is difficult to channel into a structured model. In fact, given the universal, human propensity to become nervous in front of an audience, regardless of size, musicians are also a convenience sample for psychologists, many of whom have no knowledge of (or experience in) concert performance (Mornell, 2002). Experiments have been conducted with musicians using interventions ranging from interventions such as biofeedback (Niemann, Pratt, & Maughan, 1993) to systematic desensitization (not particularly effective: Brugués, 2009 [review]; Clark, 1989; Wardle, 1969), in an attempt to reduce performance anxiety. These studies are based on varied MPA theories, which, in turn, are founded on general psychological principles, or derived from sport science models.

Three major anxiety theories that have been applied to music performance – the Inverted-U Hypothesis (or Yerkes-Dodson Law, [Yerkes & Dodson, 1908]); the Zones of Optimal Functioning model (ZOF; Gould & Tuffey, 1996); and Catastrophe Theory (Hardy & Parfitt, 1991) – all indicate that arousal and/or anxiety beyond an optimal level leads to music performance decrements. A common approach to stage fright is the categorization and rating of specific symptoms that occur during performance as either facilitating or debilitating using a modification of the Inverted-U Hypothesis. A continuum of activation is often suggested (see page 94 in this volume). Here healthy increases in arousal or anxiety necessary to complete a difficult task such as making music "under pressure" can be seen on the left side of the peak, where energy is facilitating. On the right an excessive quantity of arousal that may cause one to be "frozen with fear" is depicted. Such conceptual approaches that aim to provide explanations and perhaps predictors of musical mistakes on stage (similar to choking situations in athletic competition) are almost exclusively built on the same research that creates the basis for models in sport.

One aspect of performance that has been minimally studied in musicians is the concept of choking. One possible explanation is that a systematic study of musical mistakes has yet to be conducted. There are no known objective tools to determine what event during a performance would qualify as an experience of choking in music. Is a single wrong note a form of musical choking? Especially considering that just one missed note is not likely to ruin an entire performance in the eye (actually: ear) of the audience, this may not be the appropriate definition. In fact, the impact of the expressive quality of the piece may even shroud that single error from detection even in the strictest critic. Even an entire "botched" phrase does not mean musical "defeat" to the listener and is often without consequence, except for the critical self-judgment of the performer who has been trained to remember mistakes in order to improve.

One missed goal kick, however, may lose the game for an entire team, ultimately being labeled a choking episode, especially by the media. Based on the choking definition debate section mentioned earlier, however, it appears that sport and music researchers may encounter similar problems with the issue of whether one mistake (or minor error) may be labeled choking or under-performance. Thus, caution should be used when labeling musical performances as choking, and consideration must be given to the cognitions experienced by the performer, as choking or under-performance may be different. In transferring choking definitions to music, we suggest that choking in musicians may occur in conditions when skilled musicians are influenced by an increase in anxiety as described in Figure 1. When the perception of pressure leads to a sub-optimal performance - performance expected based on their expertise and practice - in anxiety-susceptible musicians, their experience fits the current understanding of choking. The study of MPA may benefit from further examination as such.

Since perception and interpretation play a role in the critique of a performance, subjectivity does not lend itself well to the identification of singular choking moments. Returning to the question of objective evaluation in music mentioned at the start of this chapter, we find a further complication in the determination of choking incidences: Since the reception of a musical performance does not always mirror the accuracy quotient perceived by the performer, whose judgment is considered objective for the sake of research, the audience's, the critic's, or the performer's? A musician may have a sudden increase in anxiety on stage, yet this may not be reflected in the behavioral outcome (a "near miss"). It is suggested that it may be the performer's comfort or discomfort will be reflected in their self-confidence and thus will be important going forward to the next performance. Therefore, the choking experience can be a nuanced, subjective experience of the musician, which is not necessarily audibly or visibly perceived by others, which may be difficult to objectively measure in research.

Attention-Based Styles in Music

Many musicians would be aware that any shift of attention - whether to irrelevant or relevant cues - could cause an interruption in an otherwise smooth performance. Some of the most detrimental moments are those in which attention is shifted towards movement control (Wulf, & Lewthwaite, 2010), making unconscious retrieval of motor patterns conscious to the performer. Duke et al. (2011) conducted an initial study on attentional focus in music, finding evidence that external focus (attention to the effects of a movement rather than the movements themselves) produced more accurate performances, although their participants were not professionals. In a review article, Wulf and Mornell (2008) pointed towards the advantages of an external focus both for musicians and their teachers. One explanation for this effect is that it aids the facilitation of movement automaticity (Wulf & Prinz, 2001).

To further examine this theory, a series of studies were conducted with professional musicians (Mornell & Wulf, 2014). Results indicated effects of musicians' focus of attention on the quality of their performance, both in terms of technical perfection and musical expression. When instructions were given to the musician to attend to the external effects of their actions (in this case, the musical expression), their performances were rated better by judges in a blind test or to play "as you usually do." Similarly, when musicians' attention was directed to their movements in an internal focus condition ("pay attention to the precision of their finger movements"), performance deteriorated and judges' ratings were poorer both on technical and interpretative measures than for external focus.

Distraction and Self-Focus Models in Music

In the early 20th century, psychoanalyst Sándor Ferenczi (1926) connected stage fright to narcissistic self-observation, describing the musician's attention divided on stage between the task at hand and subjective evaluation of his/her own behavior, creating a performance disturbance. In other words, it is assumed that an artist is preoccupied with his own self-image and concern about self-presentation can impair performance. This preoccupation with self may lead to attentional shifts during execution of the skilled task. Wan and Huon (2005) conducted an investigation into the distinction between distraction and self-focus theories in music in lay musicians. Mirroring Beilock and Carr (2001), Wan and Huon taught novice musicians a keyboard task under single-task, dual-task, and video-monitoring conditions, and then examined performance under low pressure and high pressure. They found that performance degradation occurred in the high-pressure, than in the low-pressure, for the single-task and dual-task groups, but performance in the high-pressure improved in the video-monitoring group. Participants who learned the task normally and under the dual-task condition were not accustomed to paying attention to the quality of their performance. Their performance quality decreased. In contrast, the performance of the video-monitoring group, which was familiar with self-monitoring conditions, did not suffer under pressure. It should be noted that specific expert and novice differences have been reported in the sport literature with novices likely to experience choking through distraction, whereas experts mostly experience choking through self-focus (e.g., Beilock, Carr, MacMahon, & Starkes, 2002). While the Wan and Huon results cannot be extrapolated to explain a choking experience in accomplished musicians or generalized to other musical disciplines (e.g., singers), they do lend first insights into the relationship between self-monitoring in practice and self-monitoring in performance.

In addition to self-monitoring, further evidence for the effects of musicians' cognitions on their performance have been identified. Using several questionnaires, including a German translation of Kenny's (2009) "Music Performance Anxiety Inventory," Peschke and von Georgi (2014) evaluated the answers of 130 musicians to determine correlations between personal beliefs and attitudes (e.g., self-confidence and deficit orientation) with performance anxiety, with the most striking and significant result being that fear of mistakes and negative judgment/ deficit orientation resulted in the highest correlation with MPA. In keeping with theories and sports research (e.g., Mesagno et al., 2012; Oudejans et al., 2011), poorer performance outcomes are likely in the presence of worry about competence.

Anticipatory anxiety (Lehrer, 1987), including the fear of mistakes, or the fear of fear, may also plague the musician. During the performance, both real and imagined errors may disrupt the musician's thoughts and confidence on stage causing a chain reaction that may result in a situation similar to choking. "Anticipating mistakes increases arousal, which further enhances access to memories of mistakes and feelings of humiliation" (Sataloff, Rosen, & Levy, 1999, p. 124). In contrast, when cognitions are rerouted from anxious thoughts back to concentration on the task at hand, performance improves. For example, anticipatory planning which includes coping strategies for stress-related performance problems allows performers to experience stage fright as facilitative (Lehrer, Goldman, & Strommen, 1990). Further evidence for the effectiveness of anticipatory strategies to reduce stress was tested in karaoke singing using cognitive reappraisal (Brooks, 2014).

Self-Presentation Models in Music

Ego threats are a major theme in a musician's life. The common saying "You're only as good as your last performance" reflects the fact that "every stage appearance has reality consequences" (Plaut, 1990, p. 60), and "captures the performer's sense that each time he performs he is seeking to establish anew his worth as a person" (Gabbard, 1983, p. 425). While identity inventories for self-complexity such as Linville's (1985; 1987) have yet to be applied to musicians, the sheer workload of practice necessary to become an expert at an instrument (10,000 hours of "deliberate practice" over 10 years mentioned previously: e.g., Ericsson 1996, 2002, 2003; Lehmann, 1997; Sloboda, Davidson, Howe, & Moore, 1996) leaves little time free to invest in additional fields of interest or talent. With identity more or less limited to that of a musician, each concert presents a potential challenge to the artist's feeling of self-worth with the associated adverse impact of stress (Linville, 1987).

Given that most musicians begin investing time and effort in practice at a very young age, they may be subject to "praise-induced disorders" (Berglas, 1990, p. 182) because they are potentially singled out and praised excessively for one salient attribute, in this case musicality. This predisposes musicians to self-image saving behavior such as self-handicapping (Higgins, Snyder & Berglas, 1990), while they "strive to preserve in fantasy the belief that they are cared for and valued as persons, not merely for what they do as producers" (Berglas, 1990, p. 182). Self-handicapping (SH) aids impression management (Kolditz & Arkin, 1982) "because of a desire to create an excuse that will protect one from the unpleasant evaluative implications of failure" (Newman & Wadas, 1997, p. 220). SH is any set of behaviors, whether conscious or unconscious, that provides an explanation for the quality of one's performance in advance of the event (Jones & Berglas, 1978). For example, when one becomes ill during the practice period before a performance, or feels compelled to limit rehearsal time because of the demands of the family, then one has an excuse for lack of sufficient preparation. This ego-protective strategy can be found not just in music, but in all fields, including sports. SH is based upon two premises: 1) one can *externalize* the outcome by blaming someone or something else in the case of failure, or 2) take personal responsibility for achievement and *internalize* success, if all goes well. Musicians experience psychological conflict because of their desire to protect their self-worth and reduce anxiety about concert outcomes on one hand, but may actually hinder their ability to invest the necessary time and effort in their concert preparation, on the other.

It could be argued that when MPA affects the artist's self-presentation, perception of pressure, and focus of attention, it may lead to a poorer performance, yet, sadly, there is little research completed about this phenomenon. Studies investigating the coping mechanisms of musicians have found that many artists are not receiving the psychological support they need from teachers or institutions (Kempinska-Welbel, 1995; Ryan & Andrews, 2009) with a proportion of musicians either dropping out of the field entirely, or turning to potentially damaging drug and alcohol use (Ryan & Andrews, 2009; Steptoe & Fidler, 1987; Welch et al., 2008).

Applying Concepts From Sport to Music Research

Considering the differences and similarities between sport and music outlined above, it can be suggested that performing artists and athletes may experience similar anxiety symptoms, social evaluative situations, and increased pressure, which may also lead to decreases in performance. Taken together, the anxiety- and attention-based models of choking encompass a range of potential contributing factors that may lead to similar decrements in performance in MPA (see Figure 1). Although trait anxiety measures have not consistently predicted choking or MPA (Lehrer, 1984), heightened state anxiety in all of its forms is found to be a factor in both athletes' and musicians' confidence, thus influencing the extent to which they experience somatic and cognitive anxiety. Within sport anxiety research, trait anxiety is positively correlated with state anxiety (e.g., Marchant, Morris, & Andersen, 1998; Williams & Krane, 1992) and positively predicts choking (Wang, Marchant, Morris, & Gibbs, 2004) in sport. Considering the similarities and the possible transfer between sport and music, this may influence both athletes' and musicians' confidence, and thus the extent to which they experience state-based somatic and cognitive anxiety.

Considering that social evaluative situations enhance anxiety levels (e.g., Mesagno et al., 2011; 2012), and that increased anxiety may lead to decreases in performance (i.e., choking in sport and stage fright in musicians), then it should follow that choking can occur in music. The subjective nature of measuring music performance, however, makes it difficult to confirm, which may lead to debate about its occurrence in music. Nevertheless, we argue that choking models from sport that include social anxiety and evaluation apprehension can be applied to music.

Given the artistic mystique that surrounds musical inspiration and performance, it has not been popular among professional musicians to openly discuss MPA, let alone choking under pressure. For both musicians and athletes alike, it is an embarrassing experience to have made mistakes in front of audiences and may never want to admit their mistakes, especially if they are a result of increased anxiety. Why would an elite performer (from either music or sport) want to risk the negative impression others will have of them if they openly admit to performance failure under pressure? Advocates of the self-presentation model of choking (e.g., Mesagno et al., 2011; 2012) would suggest that social and individual identity of the performer would hate to open himself/herself up to this public evaluation, thus, not admitting to it may be easier. In order to learn, performers have to be willing to openly admit and become aware of the reasons for performance failure. With conclusive, empirical evidence and proper coaching, however, it should be possible to convince these performers to learn from their own and others' experience. With an understanding of the mechanisms of choking as part of MPA, it is possible to amend practice and performance strategies especially with respect to mental preparation that expands identity, strengthens self-worth, reduces deficit orientation, and illuminates aspects of focus of attention in training.

The benefits of a "cross-over approach" for the researchers within these domains would be to extend the breadth of possible concepts that could be applied to help understand and alleviate performer's discomfort. Taking models most commonly used to explain choking from sport (i.e., primarily attention-based), and those used predominantly in the MPA literature (i.e., anxiety-based), and applying them to the respective other domain has not been done on a large scale in research to date. In addition, the theoretical constructs behind self-presentation, which include issues of self-identity and ego-protection through self-handicapping have yet to be studied extensively in choking- or MPA-susceptible performers. With the extensive catalog of factors explained in this chapter, we seek to provide an overview of the common ground that athletes and musicians share.

Through a deeper understanding of the mechanisms behind choking under pressure, treatments can be better targeted at the appropriate stage of an athlete's or a musician's learning. Music interventions, to date, have focused on anxiety reduction techniques (Kenny, Driscoll, & Ackermann, 2014), whereas sport interventions have primarily focused on attention-based interventions (Mesagno, et al, 2015). Irrespective of the interventions investigated thus far, future interventions that aim to reduce anxiety through deeper psychological consultation methods, rather than limited arousal regulation techniques (e.g., deep breathing, etc.) are needed for both domains, in order to reduce the distractions, negative cognitions, as well as the somatic and behavior symptoms that provide the breeding ground for poor performance. In particular, interventions based on the self-presentation model (Mesagno et al., 2011) in sport, could potentially increase retention and decrease dropout in both disciplines, as well as decrease dysfunctional coping (e.g., drug and alcohol [Chesky & Hipple, 1999]). Given that skilled performers need to be trained in the best possible way, the time has come that athletes and musicians look "over the fence" to discover new ways to see themselves as humans, and to learn to address such basic human issues such as fear, identity, and self-focus.

References

Altenmüller, E., Ioannou, C. I., & Lee, A. (2015). Apollo's curse: Neurological causes of motor impairments in musicians. *Progress in Brain Research, 217*, 89 – 106.

Anderson, J. R. (1987). Skill acquisition: Compilation of weak-method problem solutions. *Psychological Review, 94*, 192-210.

Appel, S. S. (1976). Modifying solo performance anxiety in adult pianists. *Journal of Music Therapy, XIII*(1), 2-16.

Ayres, J. (1990). Situational factors and audience anxiety. *Communication Education, 39*, 283-291.

Bar-Haim, Y., Lamy, D., Pergamin, L., Bakermans-Kranenburg, M. J., & Van IJzendoorn, M. H. (2007). Threat-related attentional bias in anxious and non-anxious individuals: A meta-analytic study. *Psychological Bulletin, 133*, 1–24.

Baumeister, R. F. (1984). Choking under pressure: Self-consciousness and paradoxical effects of incentives on skillful performance. *Journal of Personality and Social Psychology, 46*, 610–620.

Behne, K. E. (Writer). (1988). Musiker auf dem Bildschirm: Materialien für Forschung und Unterricht. Niedersächsisches Landesinstitut für Schulentwicklung und Bildung.

Beilock, S. L., & Carr, T. H. (2001). On the fragility of skilled performance: What governs choking under pressure? *Journal of Experimental Psychology: General, 130*, 701–725.

Beilock, S. L., Carr, T. H., MacMahon, C., & Starkes, J. L. (2002). When paying attention becomes counterproductive: Impact of divided versus skill-focused attention on novice and experienced performance of sensorimotor skills. *Journal of Experimental Psychology: Applied, 8*, 6–16.

Berglas, S. (1990). Self-handicapping: Etiological and diagnostic considerations. In R. L. Higgins, C. R. Snyder, & S. Berglas (Eds.), *Self-handicapping: The paradox that isn't* (pp. 151-186). New York: Plenum Press.

Bitzan, W. (2010). *Auswendig lernen und spielen. Über das Memorieren in der Musik.* Frankfurt am Main: Peter Lang Verlag.

Boucher, H., & Ryan, C.A. (2011). Performance stress and the very young musician. *Journal of Research in Music Education, 58*, 329-345.

Brandfonbrener, A. G. (1999). Performance anxiety: "Different strokes for different folks." *Medical Problems of Performing Artists, 14*(3), 101-102.

Brewer, B. W., Van Raalte, J. L., & Linder, D. E. (1993). Athletic identity: Hercules' muscles or Achilles heel? *International Journal of Sport Psychology, 24*, 237- 254.

Broadhurst, P. L. (1957). Emotionality and the Yerkes-Dodson law. *Journal of Experimental Psychology, 54*(5), 345-352.

Brodsky, W. (1996). Music performance anxiety reconceptualized: A critique of current research practices and findings. *Medical Problems of Performing Artists, 11*(3), 88-98.

Brooks, A. W. (2014). Get excited: Reappraising pre-performance anxiety as excitement. *Journal of Experimental Psychology: General, 143*(3), 1133-1158.

Brotons, M. (1994). Effects of performing conditions on music performance anxiety and performance quality. *Journal of Music Therapy, 31*, 63-81.

Brown, B. R., & Garland, H. (1971). The effects of incompetency, audience acquaintanceship, and anticipated evaluative feedback on face-saving behavior. *Journal of Experimental Social Psychology, 7*, 490-502.

Brugués, A. O. (2009). *Music performance anxiety. A review of the literature.* (Medizinischer Doktorgrad), Albert-Ludwigs-Universität, Freiburg i. Br.

Brugués, A. O. (2011). Music performance anxiety – Part 1. A review of its epidemiology. *Medical Problems of Performing Artists, 26*, 102 – 105.

Burstein, S., & Meichenbaum, D. (1979). The Work of Worrying in Children Undergoing Surgery. *Journal of Abnormal Child Psychology, 7*(2), 121-132.

Burton, D. (1988). Do anxious swimmers swim slower? Reexamining the elusive anxiety-performance relationship. *Journal of Sport & Exercise Psychology, 10*, 45-61.

Casals, P. (1970). *Joys and sorrows: Reflections by Pablo Casals as told to Albert E. Kahn.* New York: Simon & Schuster.

Chesky, K. S., & Hipple, J. (1999). Musicians' perceptions of widespread drug use among musicians. *Medical Problems of Performing Artists, 14*(4), 185-195.

Chua, A., & Rubenfeld, J. (2014). *The triple package. What really determines success.* London: Bloomsbury.

Clark, D. B. (1989). Performance-related medical and psychological disorders in instrumental musicians. *Annals of Behavioral Medicine, 11*(1), 28-34.

Clark, D. B., & Agras, W. S. (1991). The assessment and treatment of performance anxiety in musicians. *American Journal of Psychiatry, 148*(5), 598-605.

Cooper, C. L., & Wills, G. I. D. (1989). Popular musicians under pressure. *Psychology of Music, 17*, 22-36.

Cox, W. J., & Kenardy, J. (1993). Performance anxiety, social phobia, and setting effects in instrumental music students. *Journal of Anxiety Disorders, 7*(1), 49-60.

Desberg, P., & Marsh, G. D. (1988). *Controlling stagefright: Presenting yourself to audiences from one to one thousand.* Oakland, CA: New Harbinger Publications.

Dews, C. L. B., & Williams, M. S. (1989). Student musicians' personality styles, stresses, and coping patterns. *Psychology of Music, 17*, 37-47.

Duke, R. A., Cash, C. D., & Allen, S. E. (2011). Focus of attention affects performance of motor skills in music. *Journal of Research in Music Education, 20*(10), 1-12.

Elliott, C. A. (1995/1996). Race and gender as factors in judgments of musical performance. *Bulletin for the Council for Research in Music Education, 127*, 50-56.

Epstein, S. (1983). Natural healing processes of the mind: Graded stress inoculation as an inherent coping mechanism. In D. Meichenbaum (Ed.), *Stress reduction and prevention* (pp. 39-66). New York: Plenum Press.

Ericsson, K. A. (Ed.) (1996). *The road to excellence: The acquisition of expert performance in the arts and sciences, sports, and games.* Mahwah, NJ: Lawrence Erlbaum Associates.

Ericsson, K. A. (2002). Attaining excellence through deliberate practice: Insights from the study of expert performance. In M. Ferrari (Ed.), *The pursuit of excellence through education* (pp. 21-55). Mahway, NJ: Lawrence Erlbaum Associates.

Ericsson, K. A. (2003). The acquisition of expert performance as problem solving: Construction and modification of mediating mechanisms through deliberate practice. In J. E. Davidson & R. J. Sternberg (Eds.), *The psychology of problem solving* (pp. 31-83). Cambridge: Cambridge University Press.

Ericsson, K. A., Krampe, R. T., & Tesch-Romer, C. (1993). The role of deliberate practice in the acquisition of expert performance. *Psychological Review, 100*, 393-394.

Eysenck, M. W., & Calvo, M. G. (1992). Anxiety and performance: The Processing Efficiency Theory. *Cognition and Emotion, 6*, 409–434.

Eysenck, M. W., Derekshan, N., Santos, R., & Calvo, M. G. (2007). Anxiety and cognitive performance: Attention Control Theory, *Emotion, 7*, 336-353.

Fenz, W. D., & Epstein, S. (1967). Gradients of psychological arousal in parachutists as a function of an approaching jump. *Psychosomatic Medicine, XXIX*(1), 33-51.

Ferenczi, S. (1926). Stage-fright and narcissistic self-observation (J. I. Suttie, Trans.). In J. Rickman (Ed.), *Theory and technique of psychoanalysis* (Vol. 2, pp. 421-422). New York: Basic Books, Inc. (Reprinted from: Further contributions to the theory and technique of psycho-analysis, Vol. II, New York, Basic Books, Inc. 1960).

Fitts, P. M., & Posner, M. T. (1967). *Human performance.* Belmont, CA: Brooks/- Cole.

Gabbard, G. O. (1983). Further contributions to the understanding of stage fright: Narcissistic issues. *Journal of the American Psychoanalytic Association, 31*, 423-441.

Geukes, K., Mesagno, C., Hanrahan, S. J., & Kellmann, M. (2013). Activation of self-focus and self-presentation traits under private, mixed, and public pressure. *Journal of Sport & Exercise Psychology, 35*, 50-59.

Gould, D., Jackson, S. A., & Finch, L. (1993). Sources of stress in national champion figure skaters. *Journal of Sport & Exercise Psychology, 15*, 134-159.

Gould, D., & Tuffey, S. (1996). Zones of optimal functioning research: A review and critique. *Anxiety, Stress & Coping: An International Journal, 9*, 53-68.

Gucciardi, D. F., Longbottom, J., Jackson, B., & Dimmock, J. A. (2010). Experienced golfers' perspectives on choking under pressure. *Journal of Sport and Exercise Psychology, 32*, 61-83.

Halvari, H., & Gjesme, T. (1995). Trait and state anxiety before and after competitive performance. *Perceptual and Motor Skills, 81*, 1059-1074.

Hamann, D. L. (1982). An assessment of anxiety in instrumental and vocal performances. *Journal of Research in Music Education, 30*, 77-90.

Hamann, D. L. (1985). The assessment of trait-state anxiety and musical performance. *Advances in Personality Assessment, 5*, 133-155.

Hamilton, K. (2008). *After the golden age: Romantic pianism and modern performance.* New York, NY: Oxford University Press, Inc.

Hammer, S. E. (1996). The effects of guided imagery through music on state and trait anxiety. *Journal of Music Therapy, 33*(1), 47-70.

Hardy, L., & Parfitt, G. (1991). A catastrophe model of anxiety and performance. *British Journal of Psychology, 82*, 163-178.

Havas, K. (1973). *Stage fright: Its causes and cures, with special reference to violin playing.* London: Bosworth & Co. Ltd.

Hemery, D. (1986). *Sporting excellence: A study of sport's highest achievers.* Champaign, Illinois: Human Kinetics Books.

Higgins, R. L., Snyder, C. R., & Berglas, S. (Eds.). (1990). *Self-handicapping: The paradox that isn't.* New York: Plenum Press.

Hill, D. M., Hanton, S., Fleming, S., & Matthews, N. (2009). A re-examination of choking in sport. *European Journal of Sport Science, 9*(4), 203-212.

Jackson, R. C., Ashford, J. J., & Norsworthy, G. (2006). Attentional focus, dispositional reinvestment and skilled performance under pressure. *Journal of Sport & Exercise Psychology, 28*, 49–68.

Jackson, J. M., & Latané, B. (1981). All alone in front of all those people: Stage fright as a function of number and type of co-performers and audience. *Journal of Personality and Social Psychology, 40*(1), 73-85.

Jones, E. E., & Berglas, S. (1978). Control of attributions about the self through self-handicapping strategies: The appeal of alcohol and the role of underachievement. *Personality and Social Psychology Bulletin, 4*(2), 200-206.

Jones, G., & Swain, A. B. J. (1995). Predispositions to experience debilitative and facilitative anxiety in elite and nonelite performers. *The Sport Psychologist, 9*, 202-212.

Jordet, G., & Hartman, E. (2008). Avoidance motivation and choking under pressure in soccer penalty shootouts. *Journal of Sport & Exercise Psychology, 30*, 450-457.

Kane, M. J., & Engle, R. W. (2000). Working-memory capacity, proactive interference, and divided attention: Limits on long-term memory retrieval. *Journal of Experimental Psychology: Learning, Memory, and Cognition, 26*, 336-358.

Kempinska-Welbel, J. (1995). Psychological Counselling for Students and Teachers of Music Schools in Poland. In M. Manturzewska, K. Miklaszewski, A. Bialkowski, & R. Bourke (Eds.), *Psychology of Music Today. Proceedings of the International Seminar of Researchers and Lecturers in the Psychology of Music.* Warsaw: Fryderyk Chopin Academy of Music.

Kendrick, M. J., Craig, K. D., Lawson, D. M., & Davidson, P. O. (1982). Cognitive and behavioral therapy for musical-performance anxiety. *Journal of Consulting and Clinical Psychology, 50*(3), 353-362.

Kenny, D. T. (2009, January). *The factor structure of the revised Kenny Music Performance Anxiety Inventory.* Paper presented at the International Symposium on Performance Science, Auckland, New Zealand.

Kenny, D. T. (2010). The role of negative emotions in performance anxiety. In P. N. Juslin & J. A. Sloboda (Eds.) *Handbook of music and emotion: Theory, research, applications* (pp. 425–451). Oxford: Oxford University Press.

Kenny, D. T. (2011). *The psychology of music performance anxiety.* New York: Oxford University Press.

Kenny, D. T., Davis, P., & Oates, J. (2004). Music performance anxiety and occupational stress amongst opera chorus artists and their relationship with state and trait anxiety and perfectionism. *Anxiety Disorders, 18*, 757-777.

Kenny, D. T., Driscoll, T., & Ackermann, B. (2014). Psychological well-being in professional orchestral musicians in Australia: A descriptive population study. *Psychology of Music, 42*, 210-232.

Kenny, D. T., Fortune, J. M., & Ackermann, B. (2011). Predictors of music performance anxiety during skilled performance in tertiary flute players. *Psychology of Music, 41*(3), 306-328.

Kolditz, T. A., & Arkin, R. M. (1982). An impression management interpretation of the self-handicapping strategy. *Journal of Personality and Social Psychology, 43*(3), 492-502.

Leary, M. R. (1992). Self-presentational processes in exercise and sport. *Journal of Sport & Exercise Psychology, 14*, 339-351.

Leary, M. R. (2001). Social anxiety as an early warning system: A refinement and extension of the self-presentation theory of social anxiety. In S.G. Hoffman & P.M. DiBartolo (Eds.), *From social anxiety to social phobia: Multiple perspectives* (pp. 321-334). Needham Heights, MA: Allyn & Bacon.

LeBlanc, A., Jin, Y.C., Obert, M., & Siivola, C. (1997). Effects of audience on music performance anxiety. *Journal of Research in Music Education, 45*, 480-496.

Lehmann, A. C. (1997). The acquition of expertise in music: Efficiency of deliberate practice as a moderating variable in accounting for sub-expert performance. In I. Deliège & J. A. Sloboda (Eds.), *Perception and cognition of music* (pp. 161-187). East Sussex, UK: Psychology Press Ltd.

Lehrer, P. M. (1984). *The causes and cures of performance anxiety: A review of the psychological literature.* Paper presented at the Biology of music making, Denver, CO.

Lehrer, P. M. (1987). A review of the approaches to the management of tension and stage fright in music performance. *Journal of Research in Music Education, 35*(3), 143-152.

Lehrer, P. M., Goldman, N. S., & Strommen, E. F. (1990). A principal components assessment of performance anxiety among musicians. *Medical Problems of Performing Artists, 5*(1), 12-18.
Linville, P. W. (1985). Self-complexity and affective extremity: Don't put all of your eggs in one cognitive basket. *Social Cognition, 3*(1), 94-120.
Linville, P. W. (1987). Self-complexity as a cognitive buffer against stress-related illness and depression. *Journal of Personality and Social Psychology, 52*(4), 663-676.
Magill, R.A. (2011). *Motor learning and control: Concepts and applications* (9th ed.). Dubuque, IA: McGraw-Hill.
Marchant, D. B., Morris, T., & Andersen, M. B. (1998). Perceived importance of outcome as a contributing factor in competitive state anxiety. *Journal of Sport Behavior, 21*, 71-91.
Marchesani, R. (2002). From the couch to The Concert: Streisand as doctor and patient. *Psychotherapy Patient, 11*, 93-97.
Martens, R., Vealey, R. S., & Burton, D. (1990). *Competitive anxiety in sport.* Champaign, IL: Human Kinetics.
Masters, R. S. W. (1992). Knowledge, knerves, and know how: The role of explicit versus implicit knowledge in the breakdown of a complex sporting motor skill under pressure. *British Journal of Psychology, 83*, 530–541.
McPherson, G. E. (2009). The role of parent's in children's musical development. *Psychology of Music, 37*(1), 91-110.
Meichenbaum, D., & Cameron, R. (1983). Stress inoculation training. In D. Meichenbaum & M. E. Jaremko (Eds.), *Stress reduction and prevention* (pp. 115-154). New York: Plenum Press.
Mesagno, C. (2009, June). *Choking under pressure: Toward a self-presentation explanation of why athletes use self-monitoring techniques.* Paper presented at the 12th World Congress of Sport Psychology, Marrakesh, Morocco.
Mesagno, C., Geukes, K., & Larkin, P. (2015). Choking under pressure: A review of current debates, literature, and interventions. In S. Mellalieu & S. Hanton (Eds), *Contemporary Advances in Sport Psychology: A Review* (pp. 148-174). London: Routledge.
Mesagno, C., Harvey, J. T., & Janelle, C. M. (2011). Self-presentational origins of choking: Evidence from separate pressure manipulations. *Journal of Sport & Exercise Psychology, 33*, 441–459.
Mesagno, C., Harvey, J. T., & Janelle, C. M. (2012). Choking under pressure: The role of fear of negative evaluation. *Psychology of Sport and Exercise, 13*, 60–68.
Mesagno, C., & Hill, D. M. (2013). Definition of choking in sport: Re-conceptualization and debate. *International Journal of Sport Psychology, 44*, 267-277.
Mishra, J. (2010). A century of memorization pedagogy. *Journal of Historical Research in Music Education, 32*, 3-18.
Möller, H. (1999). Lampenfieber und Aufführungsängste sind nicht dasselbe! *Üben & Musizieren, 5*, 13-19.
Mornell, A. (2002). *Lampenfieber und Angst bei ausübenden Musikern* (Vol. 14). Frankfurt am Main: Peter Lang.
Mornell, A. & Wulf, G. (2014). Aufmerksamkeitsfokus auf der Bühne: Wie musikalische Fehler entstehen oder vermieden werden. In: *Angewandte Musikpsychologie – Jahrestagung der Deutschen Gesellschaft für Musikpsychologie (DGM) 2014.* Tagungsband zur Jahrestagung der DGM, p. 84.
Morris, L. W., Davis, M. A., & Hutchings, C. H. (1981). Cognitive and emotional components of anxiety: Literature review and a revised Worry-Emotionality scale. *Journal of Educational Psychology, 73*, 541-555.

Newman, L. S., & Wadas, R. F. (1997). When the stakes are higher: Self-esteem instability and self-handicapping. *Journal of Social Behavior and Personality, 12*(1), 217-233.
Nicholson, D. R., Cody, M. W., & Beck, J. G. (2015). Anxiety in musicians: On and off stage. *Psychology of Music, 43*(3), 438-449.
Nideffer, R. M. (1992). *Psyched to win.* Champaign, IL: Leisure Press.
Niemann, B. K., Pratt, R. R., & Maughan, M. L. (1993). Biofeedback training, selected coping strategies, and music relaxation interventions to reduce debilitative musical performance anxiety. *International Journal of Arts Medicine, 2*(2), 7-15.
Osborne, M.S., & Kenny, D.T. (2008). The role of sensitizing experiences in music performance anxiety in adolescent musicians. *Psychology of Music, 36*, 447-462.
Otten, M. (2009). Choking vs. clutch performance: A study of sport performance under pressure. *Journal of Sport & Exercise Psychology, 31*, 583–601.
Oudejans, R. R. D., Kuijpers, W., Kooijman, C. C., & Bakker, F. C. (2011). Thoughts and attention of athletes under pressure: Skill-focus or performance worries? *Anxiety, Stress and Coping, 24*, 59-73.
Papageorgi, I., Hallam, S., & Welch, G. F. (2007). A conceptual framework for understanding musical performance anxiety. *Research Studies in Music Education, 28*, 83-107.
Peschke, S., & von Georgi, R. (2014, August). *Die Kompetenz des Leistungsabrufes: Mentale Faktoren des Gelingens und Scheiterns von Musikern.* Paper presented at the Angewandte Musikpsychologie - Jahrestagung der Deutschen Gesellschaft für Musikpsychologie (DGM) 2014, Erlangen.
Plaut, E. A. (1990). Psychotherapy of performance anxiety. *Medical Problems of Performing Artists, 5*(1), 58-63.
Pomerantz, E. M., Grolnick, W. S., & Price, C. E. (2005). The role of parents in how children approach achievement: A dynamic process perspective. . In A. J. Elliot & C. S. Dweck (Eds.), *Handbook of Competence and Motivation* (pp. 259-278). New York: Guilford Press.
Rotella, R. J., & Lerner, J. D. (1993). Responding to competitive pressure. In R. N. Singer, M. Murphey & L. K. Tennant (Eds.), *Handbook of research on sport psychology* (pp. 528-541). New York: Macmillan.
Ryan, C., & Andrews, N. (2009). An investigation into the choral singer's experience of music performance anxiety. *Journal of Research in Music Education, 57*, 108-126.
Salmon, P. (1990). A psychological perspective on musical performance anxiety: A review of the literature. *Medical Problems of Performing Artists, 5*(1), 2-11.
Salmon, P., Schrodt, R., & Wright, J. (1989). A temporal gradient of anxiety in a stressful performance context. *Medical Problems of Performing Artists, 4*(2), 77-80.
Samson, J. (1996). *Chopin.* Oxford: Oxford University Press.
Sataloff, R. T., Rosen, D. C., & Levy, S. (1999). Medical treatment of performance anxiety: A comprehensive approach. *Medical Problems of Performing Artists, 14*(3), 121-126.
Scanlan, T. K., Babkes, M. L., & Scanlan, L. A. (2005). Participation in sport: A developmental glimpse at emotion. In J. L. Mahoney, R. W. Larson, & J. S. Eccles (Eds.), *Organized activities as contexts of development: Extracurricular activities, after-school and community programs* (pp. 108-114). Mahwah, NJ: Lawrence Erlbaum.
Scanlan, T. K., Stein, G. L., & Ravizza, K. (1991). An in depth study of former elite figure skaters: III. Sources of stress. *Journal of Sport & Exercise Psychology, 13*, 103-120.
Schlenker, B. R. (1980). *Impression management: The self-concept, social identity, and interpersonal relations.* Monterey, CA: Brooks/Cole.
Simoens, V. L., Puttonen, S., & Tervaniemi, M. (2015). Are music performance anxiety and performance boost perceived as extremes of the same continuum? *Psychology of Music, 43*(2), 171-187.

Sloboda, J. A., Davidson, J. W., Howe, M. J. A., & Moore, D. G. (1996). The role of practice in the development of performing musicians. *British Journal of Psychology, 87*, 287-309.
Spahn, C. (2012). *Lampenfieber: Handbuch für den erfolgreichen Auftritt. Grundlagen, Analyse, Maßnahmen.* Leipzig: Henschel
Spielberger, C. D. (1966). Theory and research on anxiety. In C. D. Spielberger (Ed.), *Anxiety and behavior* (pp. 3-19). New York: Academic Press.
Steptoe, A. (2001). Negative emotions in music making: The problem of performance anxiety. In P. N. Juslin & J. A. Sloboda (Eds.), *Music and emotion: Theory and research* (pp. 291-307). Oxford, UK: Oxford University Press.
Steptoe, A., & Fidler, H. (1987). Stage fright in orchestral musicians: A study of cognitive and behavioural strategies in performance anxiety. *British Journal of Psychology, 78*, 241-249.
Sternbach, D. J. (1993). Addressing stress-related illness in professional musicians. *Maryland Medical Journal, 42*(3), 283-288.
Sweeney, G. A., & Horan, J. J. (1982). Separate and combined effects of cue-controlled relaxation and cognitive restructuring in the treatment of musical performance anxiety. *Journal of Counseling Psychology, 29*(5), 486-497.
Tommasini, A. (2000, March 25, 2000). An enigmatic pianist reclaims her stardom, The Arts/ Cultural Desk. *New York Times*, p. 1.
van Kemenade, J. F., van Son, M. J., & van Heesch, N. C. (1995). Performance anxiety among professional musicians in symphonic orchestras: A self-report study. *Psychological Reports, 77*(2), 555-562.
Wan, C. Y., & Huon, G. F. (2005). Performance degradation under pressure in music: An examination of attentional processes. *Psychology of Music, 33*, 155-172.
Wang, J., Marchant, D. B., Morris, T., & Gibbs, P. (2004). Self-consciousness and trait anxiety as predictors of choking in sport. *Journal of Science and Medicine in Sport, 7*, 174-185.
Wardle, A. (1969). *Behavior modification by reciprocal inhibition of instrumental music performance anxiety.* New York: Teachers College Press.
Watson, D., & Friend, R. (1969). Measurement of social-evaluative anxiety. *Journal of Consulting and Clinical Psychology, 33*, 448-457.
Welch, G., Papageorgi, I., Creech, A., Himonides, E., Potter, J., Haddon, E., . . . Morton, F. (2008). Investigating musical performance: Performance anxiety across musical genres. *TLRP Research Briefing, 57*, 1-4.
Williams, J. M., & Krane, V. (1992). Coping styles and self-reported measures of state anxiety and self-confidence. *Journal of Applied Sport Psychology, 4*, 134-143.
Williamson, A. (1999). The value of performing from memory. *Psychology of Music, 27*, 87-95.
Wine, J. D. (1971). Test anxiety and direction of attention. *Psychological Bulletin, 76*, 92–104.
Wine, J. D. (1982). Evaluation anxiety: A cognitive-attentional construct. In H. W. Krohne & L. Laux (Eds.), *Achievement, stress, and anxiety* (pp. 207-219). New York: Hemisphere Publishing Corporation.
Woodman, T., & Hardy, L. (2001). Stress and anxiety. In R. N. Singer, H. A. Hausenblas, & C. M. Janelle (Eds.), *Handbook of Sport Psychology* (pp. 127–170). New York: Wiley.
Wulf, G. (2013). Attentional focus and motor learning: A review of 15 years. *International Review of Sport and Exercise Psychology, 6*(1), 77-104.
Wulf, G., Höß, M., & Prinz, W. (1998). Instructions for motor learning: Differential effects of internal versus external focus of attention. *Journal of Motor Behavior, 30*, 169-179.
Wulf, G., & Lewthwaite, R. (2010). Effortless motor learning? An external focus of attention enhances movement effectiveness and efficienc. In B. Bruya (Ed.), *Effortless attention: A new perspective in attention and action* (pp. 75-101). Cambridge, MA: MIT Press.

Wulf, G., McNevin, N. H., & Shea, C. H. (2001). The automaticity of complex motor skill learning as a function of attentional focus. *Quarterly Journal of Experimental Psychology 54A*, 1143-1154.
Wulf, G., & Mornell, A. (2008). Insights about practice from the perspective of motor learning: a review. *Music Performance Research, 2*, 1-25.
Wulf, G., & Prinz, W. (2001). Directing attention to movement effects enhances learning: A review. *Psychonomic Bulletin & Review, 8*(4), 648-660.
Yerkes, R. M., & Dodson, J. D. (1908). The relation of strength of stimulus to rapidity of habit-formulation. *Journal of Comparative Neurology and Psychology, 18*, 459-482.
Yoshie, M., Kanazawa, E., Kudo, K., Ohtsuki, T., & Nakawaza, K. (2011). Music performance anxiety and occupational stress among classical musicians. In J. Langan-Fox & C. L. Cooper (Eds.), *Handbook of Stress in the Occupations* (pp. 409-425). Cheltenham, UK: Edward Elgar Publishing Limited.
Yoshie, M., Shigemasu, K., Kudo, K., & Ohtsuki, T. (2009). Effects of state anxiety on music performance: Relationship between the Revised Competitive State Anxiety Inventory-2 subscales and piano performance. *Musicae Scientiae, 8*, 55-84.

The Impact of Stress and Recovery on Performance

Introduction

People who want to achieve high-level performances well know the statement "No sweet without sweat!" It expresses the accepted opinion that somebody has to invest high amounts of time, energy, and other costs (e.g., missing social events) to achieve his/her goals. Ericsson and Charness (1994) suggested that musicians, athletes, and performers in other domains must practice for 10,000 hours or 10 years to become experts in their activity. However, being in a recovered condition is necessary to reach top performance levels and to achieve training effects. "I am better this year because I train less; in other years I was already tired before the race" commented a professional cyclist on his late-season fitness levels underlining the need for balanced training and recovery periods (Süddeutsche Zeitung, October 27, 1995). Thus, research in sports has already determined that extensive training holds many pitfalls if sufficient psychological and physiological recovery is not ensured (e.g., underrecovery syndrome, overtraining; see Kellmann & Altfeld, 2014a, b). But what could musicians and performing artists learn from athletes? Are there approaches in sports that could be adapted to improve musicians' performances? In this chapter, the authors address these challenging similarities of musicians/performing artists and high performance athletes and encourage utilizing current methods used in sports to optimize performances.

Recovery From Stress (Including Exercise)

> A person can be met in a wide range of physical, cognitive and motivational states, which determine how the person will act and react to a task or a treatment in a certain situation. The recovery-stress state is an important aspect of the persons' state and is reflected in emotion, motivation, activation, and cognition as well as psychophysical symptoms and in current social activities (Kallus, 2016a, p. 27).

This definition captures that the individual stress and recovery state is a critical component in personal performing situations (Shriee & Hallé, 2011; Steenland & Deddens, 1997). While research has established that the quality and quantity of stress affect individual coping modes, the effective use of periods of recovery (e.g., breaks during the day, weekends, vacations) and the compensation of stress by enjoyable activities (e.g., exercise, social contact, sleep, reading) seems to be often overlooked by athletes, coaches, or other performers. However, to prevent critical health issues (e.g., overtraining, burnout) and performance deterioration, one approach might be the focus and optimization of recovery activities in an effort to create buffers to help individuals to better deal with the demands of their personal life.

For Kellmann (2002a), the principal of homeostasis is central to the process of recovery. Thus, the process can be defined as re-establishing the initial state of the homeostatic balance by accommodation of a deficit organism state (e.g., fatigue). However, this simplistic definition of recovery does not capture the complexity of the recovery construct. Therefore, by proposing characteristics of a general psychophysiological concept of recovery, Kellmann and Kallus (2001) tried to characterize this complex process:

- Recovery is a process in time.
- Recovery is related to the type of and duration of stress.
- Recovery depends on a reduction of, a change of, or a break from stress.
- Recovery is individually specific and depends on individual appraisal.
- Recovery ends when a psychophysical state of restored efficiency and homeostatic balance is reached.
- Recovery includes purposeful action (proactive recovery), as well as automated psychological and biological processes restoring the initial state (passive recovery).
- Recovery can be described on various levels (e.g., physiological level, psychological level, social level, sociocultural level, environmental level).
- Recovery processes can be displayed in various organismic sub-systems.
- Various subprocesses of recovery can be dissociated.
- Recovery is closely tied to boundary conditions (e.g., sleep, social contact, etc.).

Consequently, Kellmann and Kallus (2001) defined recovery as:

> an inter- and intra-individual multilevel (e.g., psychological, physiological, social) process in time for the re-establishment of performance abilities. Recovery includes an action-oriented component, and these self-initiated activities (proactive recovery) can be systematically used to optimize situational conditions to build up and refill personal resources and buffers (p. 22).

This definition was pictured by Kellmann's (2002a) scissor model describing the inter-relationship between stress-states and recovery demands. In this model, high stress levels are not detrimental as long as the person engages sufficiently in recovery activities. The basic assumption of this model is that with increased stress, increased recovery is necessary to prevent any further rise in stress. In regard to musicians, this approach implies that long, intensive practice sessions result in higher demands of recovery. However, limited resources (e.g., time) can initiate a vicious cycle. Specifically, increased stress and an inability to meet increased recovery demands can result in a person experiencing more stress. Therefore, recovery demands are defined as the quality and/or quantity of recovery activities needed to balance the current recovery-stress state. Consequently, if an athlete or musician is not able to meet recovery demands, stress accumulates, and without intervention, underrecovery symptoms (e.g., overtraining, lack of motivation, injuries, burnout) could arise.

Another important aspect concerns the conditions of adequate recovery which could disturb the re-fill process of individual energy stores. If the requirements of adequate recovery are met but the person is confronted with an emotional discussion, noise, or lack of/wrong debriefing of performance (Hogg & Kellmann, 2002), it can nevertheless result in *disturbed recovery*. Kallus and Krauth (1995) outlined on their examinations that during the regeneration pause, there is a considerable increase in the amount of sensitivity towards disturbances, annoyances, and irritations. Thus, slight interruptions in the normal situation during the break may have great effects on performance (Kellmann & Altfeld, 2014a, b; Kenttä & Hassmén, 2002).

Additionally, high performance and intensive training can stimulate adaptive processes during recovery, which not only refill resources but overcompensate to adapt for future performance. This effect requires recovery processes and in some instances constructive coping with preceding stress situations.

Musicians Need to Perform Like Sport Performers

In high performance sports, little details could impact whether you win the gold medal or drop out in the first trial. On the highest performance level, performers have usually similar technical and physical abilities. Thus, differences in the recovery state and in mental skills could be the crucial points regarding success, failure and/or injury (Sheard & Golby, 2006; Shriee & Hallé, 2011; Steenland & Deddens, 1997). These circumstances are somewhat comparable to the situation of professional musicians. Due to the increasing globalization and consequently stronger and worldwide competition, professional musicians are confronted with increasing demands on quality, technical skills, and virtuosity (Zalpour, 2013).

Gabrielsson (1999) reported in an overview of different research results that professional musicians are confronted with different feelings and situations which are able to negatively influence their recovery-stress state. Working conditions of musicians encompass a broad range of stress factors. On the emotional side performance anxiety, fear of 'others/ younger colleagues that may play better,' fear of competition, and self-doubt contribute to suboptimal recovery-stress balance. Additionally, stress conditions of the situation, respective working conditions like work overload, problems due to a working life which is often linked with irregular schedules (e.g., working in evening hours, on weekends, travelling, concert tours, etc.) as well as lack of sleep (e.g., because of travelling or sleep disturbances like difficulties in sleeping after concerts) can occur. Finally, social conditions like isolation during concert tours, 'family-work conflicts' between a professional career as musician and social life, conflicts with colleagues/a band member or members/the conductor/the management/record companies etc., job insecurity, and other issues impact the recovery-stress balance of musicians. Regarding job insecurity it should be mentioned that musicians are often self-employed and have to deal with non-permanent forms of employment, short-term contracts, employment on a project-by-project basis and absence of rights to paid holidays (Galloway, Lindley, Davies, & Scheibl, 2002; Giga, Hoel, & Cooper, 2003; Möller & Popova, 2011). In addition, Hildebrandt (2009) noted stressful working conditions (e.g., high noise level, lack of light, lack of space) as causes for occupation-related problems of professional musicians.

Shriee and Hallé (2011) explored the relationship between potential psychological risk factors and injury risk in circus artists. 47 circus artists who wanted to become Cirque du Soleil artists completed the Recovery-Stress Questionnaire for Athletes (Kellmann & Kallus, 2001, 2016) during their first two weeks of training to assess the injury risk ratio (odds ratio). The score of the scales *Injury*, *Emotional Exhaustion*, *Self-Efficacy*, and *Fatigue* were associated with an increase in injury risk (risk ratios between 1.8 and 2.8). Regarding several specific psychological aspects that are considered risk factors for injury, high injury scores had the strongest relationship. Shriee and Hallé (2011) concluded that most of the strong psychological risk factors for injuries identified in athletes also appear to be risk factors in circus artists. In line with these results Aranđelović and Ilić (2006) refer to a study of the University of Manchester (1987 as cited in Aranđelović & Ilić, 2006) which showed for the self-perception of occupational stress, that musicians belong to occupations with high stress levels (scale from 0 to 10, examples: miner 8.3, police officer 7.7, airline pilot 7.5, actor 7.2, ambulance personnel 6.3, firefighter 6.3, musician 6.3, teacher 6.2).

Moreover, there are also similarities between professional musicians and high performance athletes regarding the frequency of competitions/concerts. Norris and Smith (2002) mentioned that the frequency of competitions structures the overall schedule (e.g., recovery times, high intensity training periods) of an athlete's season. Thereby, Kellmann (2002b) distinguished three types of performing density (low, middle, and high density of performing) having independent demands on individual training and recovery periods.

In sports with low density of performing (e.g., rowing, cycling, track and field), athletes are usually faced with three major competitions (e.g., National, World and European Championships, Olympics). Consequently, the periodization (e.g., preparation competitions) is planned around the major events (Norris & Smith, 2002). However, the importance of the World Championships or Olympics is topped by the trials to qualify for those events. These might be comparable to professional artists preparing for decisive audition. Consequently, the musicians as well as athletes are confronted with the question whether there is enough time between the qualification rounds and the final competition to go through a complete or partial periodization process.

In medium density performance sports, athletes usually have a full year competition schedule (e.g., basketball, soccer, tennis) with weekly games or competitions. Additionally, often national cup and international championship games are scheduled if the athlete or team qualifies. For example, the German first division basketball team Alba Berlin played 32 games in 100 days in the 2014/2015 season. This is an average of one game every 3.1 days, including long travel times and preparation. Consequently, physical and mental recovery is often incomplete and might impact subsequent performances negatively. These circumstances are comparable to musicians playing concerts every third evening in different cities. In a study focusing on the stress and recovery of professional ball room dancers, dancing in several competitions in different cities and countries during their season, high levels of stress were observed in the male participants (Liiv, Jurimae, Klonova, & Cicchella, 2013). Consequently, the time between the games/competitions/concerts for these performers should be focused on regenerative training and replenishing energy stores, not on high training load or high intensity training. The cornerstones of fitness and technique are usually set before the season starts or during breaks (e.g., winter break).

High density performers compete more often than in the other two types. They perform mostly one or two competitions/games/shows a day. For example, the athletes of the Cirque du Soleil have approximately 400 shows a year (often twice a day). Because of this load of situations where to perform on individual highest level, special demands regarding training structure and recovery periods are required to prevent underrecovery and overtraining, as well as lack of motivation. Thus, maintaining rather than peaking performance is the key element. Regarding this, it is not the goal that the athletes achieve 100% of their optimal performance every show but that they obtain a minimum level that enables the performance on a high level over a long period of time. Consequently, 95% of the maximum performance is aimed during the task to achieve optimal performances in decisive shows. Therefore, the ideal surroundings and conditions need to be served to optimize the recovery process and to ensure the minimum performance level.

In summary, musicians are confronted with several circumstances affecting individual recovery state and performance ability. Based on this, professional musicians need to be aware of the costs of inadequate regeneration periods. Therefore, beside breaks and changes in demands, which are needed to be planned to prevent an overload, to monitor the individual recovery state seems to be sufficient (Kellmann, 2010). In sports, athletes, coaches, and researchers already realized that enhanced recovery allows practicing more and improves technique quality and efficiency. For that reason, monitoring instruments are used in sports to assess the individual's mood, athletes' need for recovery, and current life circumstances affecting recovery states (e.g., Kenttä, Hassmén, & Raglin, 2006).

Parallels between top performance in music and sports suggest similar strategies to cope with the broad range of stressors and to keep stress and recovery in balance. "Art in Motion" indicates that musicians might profit a lot by adapting approaches from high performance sports. However, there are still many areas which call for improvements. Weigelt and Stöckel (2009) used the term 'Finger Athletes' to indicate similarities, but even fingers are not property warmed up before performance in many instances. The opposite is true for many situations, as peripheral blood flow is reduced by noise – also when the noise is part of the (musical) parts, which is performed by the colleague nearby in the orchestra. From a sport- or bio-psychological perspective the high incidence of upper extremity disorders and other stress related symptoms (like tinnitus, metabolic dysregulation, etc.) is partly due to the fact that basic rules of recovery on how to integrate the body into high performance are violated systematically in musicians.

The same applies to mental training. While the musical score is perfectly represented mentally in very high performance musicians, their own body, the body feeling and body movements, the smell, the taste, the room acoustics, etc. are hardly integrated into the mental training steps in musicians. Finally, in high performance sport competitions, training, recovery, and performance schedules are guided by coaches and trainers. For musicians, only teachers, conductors, and managers are available, which leaves the important functions of balancing stress and recovery to the musicians. Only in few instances the orchestra is a team, and one often needs to qualify – like in sports – to become a team member. Thus, engaging individually monitoring mechanisms into the work-rest schedule of musicians could be a health and performance promoting enterprise.

While the stress-health-axis is especially as well as theoretically well documented (Hellhammer, Stone, Hellhammer, & Broderick, 2010; Picard, Juster, & McEwen, 2014) specific etiological models (like for upper extremity disorders) or disorders due to missing recovery and long working hours (e.g., sleep disorders) needed more scientific effort (Gaisbachgrabner, 2014).

Assessing and Monitoring the Recovery-Stress State

The Recovery-Stress Questionnaire (RESTQ-Basic, Kallus, 2016b) can be used as assessment tool to measure the current recovery-stress state of a person. For this purpose the frequency of different stress and recovery-related states and activities in *the past (3) days/nights* are measured with 12 scales [standard time frame of the RESTQ-Basic; values ranging from 0 (*never*) to 6 (*always*); see Kallus, 2016b for more details]. The 12 different scales can be grouped into the two basic dimensions *Overall Recovery* (five scales: *Success, Social Recovery, Physical Recovery, General Well-Being, Sleep Quality*) and *Overall Stress* (seven scales). Furthermore, the dimension *Overall Stress* can be subdivided into the two sub-dimensions: *Social-Emotional Stress* (three scales: *General Stress, Emotional Stress, Social Stress*) and *Performance(-Related) Stress* (four scales: *Conflicts/Pressure, Fatigue, Lack of Energy, Physical Complaints*). The standard version of RESTQ-Basic consists of 48 items (RESTQ-Basic-48). Beneath the standard version a short version with a reduced number of items (24 items; RESTQ-Basic-24) can be used (Kallus, 2016b).

To assess stress in the workplace the Recovery-Stress Questionnaire for Work (RESTQ-Work; Jiménez, Dunkl, & Kallus, 2016) can be used as assessment tool. The RESTQ-Work combines the current recovery-stress state of a person (RESTQ-Basic) and the concept of work-specific resources and recovery. The work-specific part consists of 14 additional scales that can be grouped into four sub-dimensions: *Loss of Meaning/Burnout* (five scales: *Emotional Exhaustion, Loss of Control, Loss of Meaning, Conflicts with Colleagues, Spillover*), *Leisure/Breaks* (four scales: *Undisturbed Leisure Time, Efficient Breaks, Leisure, Undisturbed Breaks*), *Psychosocial Recovery* (two scales: *Social Support Friends, Social Support Colleagues*), and *Work-Related Recovery* (three scales: *Participation, Personal Growth Factors, Action Latitude*). The RESTQ-Work is available in three different versions: a short version with 27 items (RESTQ-Work-27) can be used for screening and two longer versions with 55 (RESTQ-Work-55) and 92 items (RESTQ-Work-92) can be used for comprehensive feedbacks.

Example for the Application of the RESTQ-WORK

Musculoskeletal and upper extremities disorders can be seen as common work-related health problems of professional musicians (Zaza & Farewell, 1997). Improper technique, repetitive movements, and ergonomic aspects cannot be seen as only causes for these disorders. Also, poor recovery, psychosocial factors and high job demands can have an impact on their occurrence (Bongers, Kremer, & ter Laak, 2002). If this hypothesis fits, professional musicians should score worse regarding stress and recovery than people employed in other professions. To verify this, the RESTQ-Work was used to check the recovery-stress state in a group of professional musicians of a symphonic orchestra. Data were compared with a representative sample of fully employed persons to check specific areas of a recovery-stress imbalance in musicians.

Method

Participants

In total, 22 professional musicians of the Opera Orchestra of the city of Graz (Austria) participated in the study: 9 women and 13 men between 24 and 56 years old with an average age of 38.43 years ($SD = 7.81$).

Measures and Procedure

Between May and June 2007 two programs to prevent work-related upper extremity disorders in musicians were introduced within the orchestra (see also Herwelly, 2007; Herwelly & Kallus, 2009). The data presented in this article were collected before the intervention period (baseline condition) and will reflect the existing situation respectively the current recovery-stress state measured with RESTQ-Work-92 of professional musicians of a symphonic orchestra.

Comparative Sample and Statistical Analyses

One-sample *t*-tests were used to compare the recovery-stress state of professional musicians of a symphonic orchestra with the recovery-stress state of employees of other professions. The comparative sample consists of 682 Austrian employees of various sectors with an average age of 42.00 years ($SD = 11.40$). 41.60% of them were male, 58.40% female. The sample is a pooled sample. Data of four different research studies which were independently conducted in different companies between 2007 and 2008 were pooled (Jiménez et al., 2016). The analyses were based on a significance level of 5%.

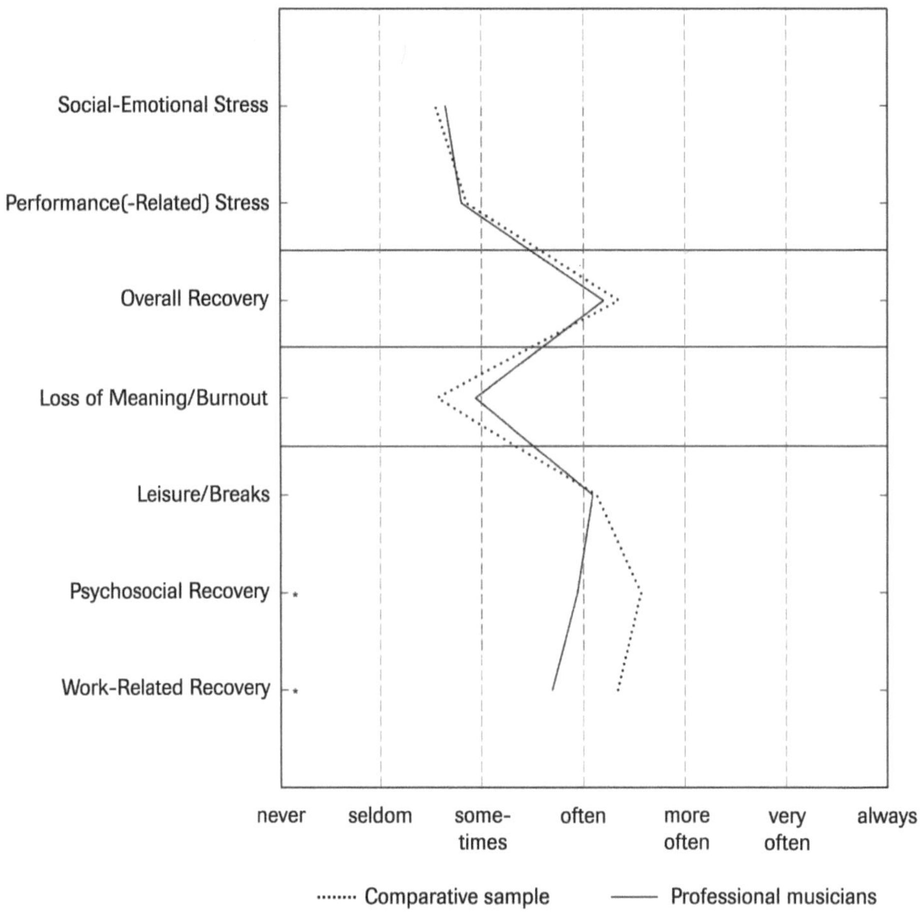

Figure 1: Comparison of the recovery-stress state of professional musicians and employees of other professions sample (sub-dimensions of RESTQ-Work-92) [p-value: *: $p < .05$]

Results

Due to an explorative approach the results were calculated on sub-dimensional basis as well as on the basis of the scales of the RESTQ-Work-92. This approach was chosen to gain information about possible differences of the recovery-stress state between professional musicians and employees of other professions.

Sub-Dimensions

The results of the one-sample *t*-tests to compare the recovery-stress state of professional musicians of a symphonic orchestra with the recovery-stress state of employees of other professions are shown in Figure 1.

Significant differences can be shown for work-specific recovery dimensions (*Psychosocial Recovery*, $t(20) = -4.983$, $p < .0001$, and *Work-Related Recovery*, $t(20) = -2.980$, $p = .007$). All other effects do not reach the level of significance.

Scales

A more detailed analysis for the scales is shown in Table 1 on p. 71. Significant differences between professional musicians and employees of other professions can be shown for some scales in three out of four sub-dimensions of the work-related part of the RESTQ-Work-92, *Loss of Meaning/Burnout*, *Psychosocial Recovery* and *Work-Related Recovery* as well as for *Success* of the recovery scales of the RESTQ-Basic.

Discussion

The explorative approach described in the result section allows a differentiation between results on two different levels: sub-dimensions as well as scales of RESTQ-Work-92. This approach was chosen to gain as much information about possible differences of the recovery-stress state between professional musicians and employees of other professions, which can be seen as successful approach for the presented study. Whereas analyses on a sub-dimensional basis showed significant differences for the two sub-dimensions *Psychosocial Recovery* and *Work-Related Recovery*, analyses taking the scales into account result in more sophisticated findings representing differences of the recovery-stress state between professional musicians and employees of other professions.

The results show that professional musicians report significantly more *Conflicts with Colleagues* and *Spillover* than employees of other professions. Following Varhama and Björkqvist (2004), the experience of conflicts with colleagues at work is strongly correlated with burnout. Additionally, spillover is associated with burnout too. Research results showed a linkage of 'work-family conflict' respectively; a spillover of work related problems to private life and emotional exhaustion of employees (Lambert &Hogan, 2010; Rupert, Stevanovic & Hunley, 2009). Schuppert (2012) notes that repressed conflicts cause stress in work environment. Orchestras can be seen as structures with many excellent individuals that have to perform as one team with high-quality. Therefore, it is essential to introduce good communication structures and conflict management. The results regarding the scale *Social Support Colleagues,* one of the two scales of the sub-dimension *Psychosocial Recovery* point in the same direction: professional musicians report significantly fewer social supports of their colleagues. Following Jiménez et al. (2016, p. 160), "[...] social support is one of the most effective stress buffers [...]," is strongly negatively related to exhaustion (Halbesleben, 2006) and has positive effects on coping with work demands (Otero-Lopéz, Mariño, & Bolanño, 2008).

Furthermore, professional musicians report significantly fewer pleasures at work, less creativity (scale *Success*), as well as a tendency towards significance for fewer frequent pleasurable social contacts and change combined with relaxation and amusement (scale *Social Recovery*). These results are in line with the results of other research groups and may be caused by the 'normal' working life of professional musicians which is often linked with irregular schedules (e.g., working in evening hours, on weekends, travelling, concert tours ...), unsociable hours, and changing locations of work (Gabrielsson, 1999; Galloway et al., 2002; Giga et al., 2003). Gabrielsson (1999) also notes that professional musicians are confronted with family-work conflicts, and conflicts between a professional career as musician and social life/isolation during concert tours. The results of a study conducted by Hamilton and Kella (1992) point in the same direction: matched with norm values, professional musicians (violinists and violists) scored above the 60th percentile for stress due to their physical environment (irregular work schedules, personal isolation, etc.) on the Measure of Occupational Stress, Strain and Coping (Osipow & Spokane, 1983).

Moreover, the results show that professional musicians experience significantly less *Work-Related Recovery* and fewer resources (scales *Participation, Personal Growth Factor, Action Latitude*) than employees of the comparative sample. *Work-Related Recovery* also includes aspects of

Table 1: Comparison of the recovery-stress state of professional musicians and employees of other professions - scales

		Comparative sample		Professional musicians			
Subdimensions	Scale	*M*	*SD*	*M*	*SD*	*t*(20)	*p*
Social-Emotional Stress	General Stress	1.37	1.04	1.35	1.16	-.051	.960
	Emotional Stress	1.55	.83	1.63	1.00	.371	.714
	Social Stress	1.70	.93	1.94	1.11	.991	.334
Perfomance (-Related) Stress	Conflicts/Pressure	2.10	.93	2.21	1.19	.411	.686
	Fatigue	2.05	1.07	2.00	1.20	-.207	.838
	Lack of Energy	1.65	.83	1.66	1.11	.003	.997
	Physical Complaints	1.61	1.00	1.34	1.17	-1.065	.300
Overall Recovery	Success	2.91	.92	2.38	.97	-2.521	.020
	Social Recovery	3.02	1.08	2.56	1.13	-1.865	.077
	Physical Recovery	3.22	1.09	3.33	1.21	.428	.673
	General Well-being	3.59	1.11	3.48	1.14	-.456	.653
	Sleep Quality	4.04	1.24	4.27	.95	1.134	.270
Loss of Meaning/ Burnout	Emotional Exhaustion	1.71	1.16	1.48	1.32	-.812	.426
	Loss of Control	1.59	1.06	2.15	1.60	1.608	.124
	Loss of Meaning	1.31	1.09	1.57	1.49	.803	.431
	Conflicts with Colleagues	1.24	.92	1.86	1.37	2.082	.050
	Spillover	1.97	1.23	2.62	1.41	2.107	.048
Leisure/ Breaks	Undisturbed Leisure Time	3.36	1.15	3.14	1.46	-.684	.502
	Efficient Breaks	3.01	1.37	3.05	1.49	.116	.909
	Leisure	3.08	1.29	3.06	1.38	-.055	.957
	Undisturbed Breaks	3.07	1.48	3.10	1.45	.080	.937
Psychosocial Recovery	Social Support Friends	3.65	1.45	3.48	1.36	-.588	.563
	Social Support Coleagues	3.49	1.16	2.80	.79	-4.041	.001
Work-Related Recovery	Participation	2.98	1.14	2.29	.94	-3.386	.003
	Personal Growth Factors	3.09	1.06	2.56	.96	-2.556	.019
	Action Latitude	3.94	1.33	3.26	1.25	-2.358	.030

control at work like autonomy or independent decision options which are possibly incompatible with the work requirements of a professional musician of an orchestra. For example, it would not be possible to make autonomous decisions which piece of music a member of an orchestra will play, or how he or she will interpret them, etc. (see also Möller & Popova, 2011). Pegelhoff (2010) postulates that opportunities of participation in decision-making can improve job satisfaction, employees' motivation and working climate in cultural institutions. Additionally, Schuppert (2012) suggests that an increase of job autonomy and latitude within orchestras can increase intrinsic motivation of professional orchestral musicians. Also, personal growth factors and promotion prospects should be improved. Note that the results indicate only few areas of increased stress but a broad range of missing resources and a lack of recovery in orchestra musicians. While stress may be difficult to change – resources and recovery should be in the focus of health and performance promoting measures for orchestra musicians. In addition, 'RecoveryChecks' (Institut für Begleitforschung, 2015) might be valuable to prevent stress related disorders and burnout.

To sum it up, professional musicians may benefit from approaches in sport, especially from research regarding training strategies that include psychological as well as physiological recovery. To achieve top performances, musicians and athletes have to cope with a broad range of stressors. Therefore, musicians may copy and/or adapt well approved athletes' strategies to keep stress and recovery in balance.

References

Aranđelović, M., & Ilić, I. (2006). Stress in workplace - possible prevention. *Facta Universitatis, Series: Medicine and Biology, 13*(3), 139-144.

Bongers, P. M., Kremer, A.-M., & ter Laak, J. (2002). Are psychosocial factors, risk factors for symptoms and signs of the shoulder, elbow, or hand/wrist?: A review of the epidemiological literature. *American Journal of Industrial Medicine, 41*(5), 315-342.

Ericsson, K. A., & Charness, N. (1994). Expert performance: Its structure and acquisition. *American Psychologist, 49*, 725-747.

Gabrielsson, A. (1999). The performance of music. In D. Deutsch (Ed.), *The psychology of music* (2nd edition, pp. 501-602). San Diego: Academic Press.

Galloway, S., Lindley, R., Davies, R., & Scheibl, F. (2002). *A balancing act: Artists' labour markets and the tax and benefit systems.* Research Report 29 - Study carried out by the Warwick Institute for Employment Research and Centre for Educational Development, Appraisal and Research University of Warwick.

Gaisbachgrabner, K. (2014). *Empirische Untersuchungen zu langen Arbeitszeiten und dem Hypothalamus-Hypophysen-Schilddrüsen-System im Mehrebenenansatz* [Long working hours and the hypothalamic-pituitary-thyroid system]. Unpublished Doctoral Dissertation, University of Graz, Austria.

Giga, S., Hoel, H., & Cooper, C. L.(2003). *Violence and stress at work in performing arts and in journalism.* Sectoral activities programme. Working Paper. International Labour Office, Geneve.

Halbesleben, J. R. B. (2006). Sources of social support and burnout: A meta-analytic test of the conservation of resources model. *Journal of Applied Psychology, 91*, 1134-1145.

Hamilton, L., & Kella, J. J. (1992). Personality and occupational stress in elite performers. *Paper presented at the Annual Convention of the American Psychological Association* (100th, Washington, DC, August 14-18, 1992), 12pp. Retrieved from http://files.eric.ed.gov/fulltext/ED353502.pdf

Hellhammer, D. H., Stone, A. A., Hellhammer, J., & Broderick, J. (2010). Measuring stress. In G. F. Koob, M. Le Moal, & R. F. Thompson (Eds.), *Encyclopedia of behavioral neuroscience* (pp. 186-191). Oxford: Academic Press.

Herwelly, S. (2007). *Gesundheitsförderung im Berufsorchester: Vergleich zweier Interventionen gegen berufsbedingte Beschwerden des oberen Bewegungsapparates* [Health promotion in the professional orchestra: Comparison of two interventions against work-related upper extremity disorders]. Unpublished Diploma Thesis, University of Graz, Austria.

Herwelly, S., & Kallus, K.W. (2009). Injury prevention in the orchestra. In A. Mornell (Ed.), *Art in motion. Musical & athletic motor learning & performance* (pp. 223-225). Frankfurt, Germany: Peter Lang GmbH.

Hildebrandt, H. (2009). Teaching music physiology and motor learning processes at a university: Experience and evaluation. In A. Mornell (Ed.), *Art in motion. Musical and athletic motor learning and performance* (pp. 191-222). Frankfurt: Peter Lang GmbH.

Hogg, J., & Kellmann, M. (2002). Debriefing im Leistungssport [Debriefing in elite sports]. *Psychologie und Sport, 9*, 90-96.

Institut für Begleitforschung (2015). RecoveryCheck. Retrieved on 22.06.2015 from http://www.begleitforschung.de/de/products/recovery.html

Jiménez, P., Dunkl, A., & Kallus, K. W. (2016). Recovery-Stress Questionnaire for Work. In K. W. Kallus & M. Kellmann (Eds.), *The Recovery-Stress Questionnaires: User Manual* (pp. 158–187). Frankfurt am Main: Pearson Assessment & Information GmbH.

Kallus, K. W. (2016a). Stress and recovery: An overview. In K. W. Kallus & M. Kellmann (Eds.), *The Recovery-Stress Questionnaires: User Manual* (pp. 27–48). Frankfurt am Main: Pearson Assessment & Information GmbH.

Kallus, K. W. (2016b). RESTQ-Basic: The General Version of the RESTQ. In K. W. Kallus & M. Kellmann (Eds.), *The Recovery-Stress Questionnaires: User Manual* (pp. 49-85). Frankfurt am Main: Pearson Assessment & Information GmbH.

Kallus, K. W., & Krauth, J. (1995). Nichtparametrische Verfahren zum Nachweis emotionaler Reaktionen [Non-parametric methods for the prove of emotional reactions]. In G. Debus, G. Erdmann, & K. W. Kallus (Eds.), *Biopsychologie von Streß und emotionalen Reaktionen* (pp. 23-43). Göttingen: Hogrefe.

Kellmann, M. (2002a). Underrecovery and overtraining: Different concepts - Similar impact? In M. Kellmann (Ed.), *Enhancing recovery: Preventing underperformance in athletes* (pp. 3-24). Champaign, IL: Human Kinetics.

Kellmann, M. (2002b). Current status and directions of recovery research. In M. Kellmann (Ed.), *Enhancing recovery: Preventing underperformance in athletes* (pp. 219-229). Champaign, IL: Human Kinetics.

Kellmann, M. (2010). Preventing overtraining in athletes in high-intensity sports and stress/recovery monitoring. *Scandinavian Journal of Medicine & Science in Sports, 20*, 95-102.

Kellmann, M., & Altfeld, S. (2014a). Overtraining syndrome. In R. C. Eklund & G. Tenenbaum (Eds.), *Encyclopedia of sport and exercise psychology* (pp. 510–512). Thousand Oaks, CA: Sage.

Kellmann, M., & Altfeld, S. (2014b). Underrecovery syndrome. In R. C. Eklund & G. Tenenbaum (Eds.), *Encyclopedia of sport and exercise psychology* (pp. 773-775). Thousand Oaks, CA: Sage.

Kellmann, M., & Kallus, K. W. (2001). *The Recovery-Stress Questionnaire for Athletes: User manual.* Champaign, IL: Human Kinetics.

Kellmann, M., & Kallus, K. W. (2016). The Recovery-Stress Questionnaire for Athletes. In K. W. Kallus & M. Kellmann (Eds.), *The Recovery-Stress Questionnaires: User manual* (pp. 86-131). Frankfurt am Main: Pearson Assessment & Information GmbH.

Winter, G., & Hassmén, P. (2002). Underrecovery and overtraining: A conceptual model. In M. Kellmann (Ed.), *Enhancing recovery: Preventing underperformance in athletes* (pp. 57-79). Champaign, IL: Human Kinetics.

Kenttä, G., Hassmén, P., & Raglin, J. S. (2006). Mood state monitoring of training and recovery in elite kayakers. *European Journal of Sport Science, 6*, 245-253.

Lambert, E. G., & Hogan, N. L. (2010). Work-family conflict and job burnout among correctional staff. *Psychological Reports, 106*, 19-26.

Liiv, H., Jurimae, T., Klonova, A., & Cicchella, A. (2013). Performance and recovery: Stress profiles in professional ballroom dancers. *Medical Problems of Performing Artists, 28*(2), 65-69.

Möller, H., & Popova, D. (2011). Der Hürdenlauf zum Orchestermusiker. Sozialisation, Ausbildung und Berufseinstieg [The hurdles to the orchestra musicians. Socialization, education, and career]. *Musikphysiologie und Musikermedizin, 18*, 39-46.

Norris, S. R., & Smith, D. J. (2002). Planning, periodization, and sequencing of training and competition: The rationale for a competently planned, optimally executed training and competition program, supported by a multidisciplinary team. In M. Kellmann (Ed.), *Enhancing recovery: Preventing underperformance in athletes* (pp. 121-141). Champaign, IL: Human Kinetics.

Otero-Lopéz, J. M., Mariño, M. J. S., & Bolanño, C. C. (2008).An integrating approach to the study of burnout in university professors. *Psicothema, 20*, 766-772.

Osipow, S. H., &Spokane A. R. (1983). *A manual for measures of occupational stress, strain, and coping*. Columbus, Ohio: Marathon Consulting and Press.
Pegelhoff, R. (2010). Mitgestaltung als Motivationsanreiz. Positive Aspekte einer mitarbeiterorientierten Unternehmensführung [Active participation as a motivational stimulus. Positive aspects of a personnel-centered management]. *das Orchester, 12*, 20.
Picard, M., Juster, R. P., & McEwen, B. S. (2014). Mitochondrial allostatic load puts the gluc back in glucocorticoids. *Nature Reviews. Endocrinology, 10*(5), 303-310.
Rupert, P., Stevanovic, P., & Hunley, H. A. (2009). Work-family conflict and burnout among practicing psychologists. *Professional Psychology: Research and Practice, 40*, 54-61.
Schuppert, M. (2012). Kreativität und Motivation: Gesundheitliche Ressource im Orchester [Creativity and motivation: Health resource in the orchestra]. *Musikphysiologie und Musikermedizin, 19*(1), 9-15.
Sheard, M., & Golby, J. (2006). Effect of a psychological skills training program on swimming performance and positive psychological development. *International Journal of Sport and Exercise Psychology, 4*, 149-169.
Shriee, I., & Hallé, M. (2011). Psychological predictors of injuries in circus artists: An exploratory study. *British Journal of Sport Medicine, 45*, 433-436.
Steenland, K., & Deddens, J. A. (1997). Effect of travel and rest on performance of professional basketball players. *Sleep, 20*, 366-369.
Varhama, L. M., & Björkqvist, K. (2004). Conflicts, workplace bullying and burnout problems among municipal employees. *Psychological Reports, 94*, 1116-1124.
Weigelt, M., & Stöckel, T. (2009). Lateral Preferences and Brain Asymmetries: Common Challenges for Sports and Finger Athletes. In A. Mornell (Ed.), *Art in motion. Musical & athletic motor learning & performance* (pp. 37-39). Frankfurt, Germany: Peter Lang GmbH.
Zalpour, C. (2013). Musikergesundheit als Aufgabe einer spezialisierten Musiker-Physiotherapie [Musician's health as task of a specialized musicians physiotherapy]. ÖFAM (Österreichisches Forum Arbeitsmedizin*), 2/2013*, 17-28.
Zaza, C., & Farewell, V. T. (1997). Musicians' playing-related musculoskeletal disorders: An examination of risk factors. *American Journal of Industrial Medicine, 32*, 292-300.

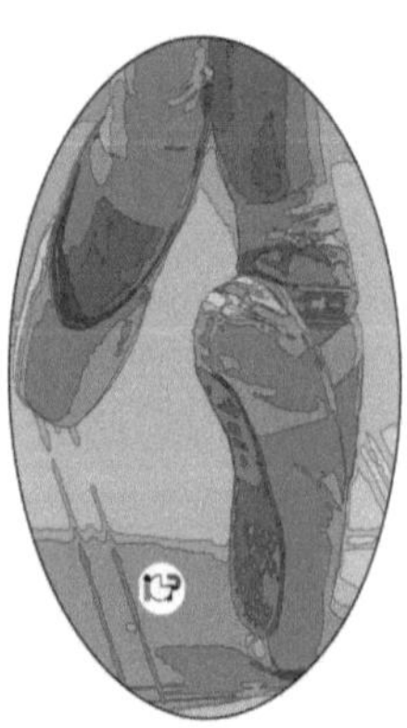

L. Aufegger, D. Wasley, A. Williamon

Facing the Music: Investigating the Psychophysiology of Musical Performance

Musicians often face intense cardiovascular and emotional demands while performing on stage. While these are commonly experienced physically by musicians during performance, relatively little systematic research has examined how they arise and dissipate in the periods before and after a stressful event, as well as their relationship to key psychological parameters that determine music performance anxiety. The first section of this chapter reviews the most recent research on applied performance physiology, particularly focusing on links with music performance anxiety. The second examines the possibilities of collecting, analyzing, and interpreting cardiovascular features, such as heart rate variability (HRV), in greater detail. Methodological considerations are raised in terms of time versus frequency domain analyses, while also offering new approaches to understanding biosignals. The last section considers new investigations of the type, frequency, and duration of physical interventions designed to optimize performance outcomes.

Applied Physiology Research in Musical Performance

Musicians' performance quality on stage relies on many intra-individual and inter-individual factors. While some, such as the amount of practice or the difficulty of the piece can be reasonably controlled, others, including the type of situation, are subject to greater variability, effectively making each performance a unique experience. Over the last four decades, much research has focused on the debilitating psychological responses associated with music performance anxiety (MPA). MPA involves cognitive, physical, or behavioral symptoms that can be evident in musicians through, for instance, memory failure, maladaptive thoughts, tremor, or failures of technique, all of which may impact on performance quality. These factors can also be experienced in combination, sometimes occurring hours, days, or weeks prior to an event. If they are reinforced through a lack of sufficient adaptive coping or vulnerable predispositions (e.g., high levels of trait anxiety), performers may experience helplessness, impaired performances, or poor health, eventually giving up and choosing a way of life with less public exposure (Kenny, 2011).

Despite the attention given to MPA in the literature, relatively little research has been undertaken to understand the wider context of MPA, namely *performance stress*, and, in particular, the physiological stress responses to musical performance. MPA is usually described as the feeling of worry and uncertainty towards a particular musical performance (and which may involve the experience physiological and behavioral symptoms). Unlike MPA, performance stress is the body's reaction to any kind of performance, and is characterized by disturbances of an organism which require physiological, but also psychological and behavioral actions to regain stability (Lazarus & Folkman, 1984). This may involve altered cardiac function including increased heart rate (HR), blood pressure, elevated breathing, or the release of stress hormones, such as cortisol. The few studies that have addressed physiological responses in musicians to performances have mainly addressed mean HR as an absolute value or as a change from baseline, expressed in beats per minute.

Cardiac function (and thus HR) is governed by the activation of a complex neuronal network globally called the central nervous system (CNS). The CNS consists of the brain and spinal cord, which transmits impulses through nerve cells and synapses that are connected throughout the body. The information transfer of the CNS is further supported by the peripheral nervous system, a vast network of mainly sensory receptors, which is divided into the somatic nervous system and the autonomic nervous system (ANS). While the somatic nervous system provides information to and from the joints, skeletal muscles, bones, and skin, the ANS controls internal organs, such those of the cardiovascular (heart), respiratory (lungs), and endocrinological (hormones) systems. The ANS further consists of two branches, the sympathetic nervous system (SNS) and the parasympathetic nervous system (PNS). Both the SNS and PNS affect changes in HR quickly, observable within five beats (Sarafino, 1996).

Studies have shown that musicians experience increased HR before (Craske & Craig, 1984; Abel & Larkin, 1990) and during performances (Brotons, 1994; Fredrikson & Gunnarson, 1992; Yoshie et al., 2009). In all cases, HR in the performance situation was significantly different from baseline and low-stress or rehearsal situations, and these differences typically occurred alongside elevated feelings of anxiety. Craske and Craig (1984) examined responses before and during musical performance. They asked musicians to perform for 3-5 minutes alone and in front of an audience of five judges while HR was collected 100, 40, and 10 seconds prior to and 120, 140, 260 and 400 seconds after the first note was played. Psychological stress responses were investigated by assessing musicians' degree of state anxiety and distress prior to the performances (Spielberger et al., 1983; Wolpe, 1969). The results showed significantly higher HR, state anxiety, and distress during the public concert and peak mean HR during the pre-performance period. Abel and Larkin (1990) asked musicians to perform in front of a jury where the pre-performance time was divided into two anticipatory periods: a warm-up period (5 minutes) and the time between the final call and the performance on stage (2 minutes). Psychological reactivity measures included state anxiety and a quick self-report of their confidence (Spielberger et al., 1983; Appel, 1974). Mean HR was taken 12 times per minute during a baseline measurement for 6 minutes, as well as for 5 minutes during a warm-up period and 2 minutes prior to going on stage. The results demonstrated significantly higher levels of state anxiety in the anticipatory periods compared with the baseline measurement, while there was no difference in their perceived level of confidence. Mean HR was significantly increased during the anticipatory periods compared with the baseline measurement, with the time just prior to going on stage exhibiting peak HR. More recently, Yoshi et al. (2009) asked musicians to perform a 5-8 minute piece from memory during a rehearsal and during a competition. HR was measured using a wristwatch monitor, while a visual analogue anxiety scale (Cella & Perry, 1986) was completed before each condition. Results revealed that musicians' mean HR and state anxiety were significantly higher prior to the jury condition.

Based on the studies available, it is difficult to draw conclusions about the timing and the degree of musicians' cardiovascular reactivity to acute psychosocial stress for at least two methodological reasons. Firstly, the studies conducted reported mean HR rather than the variability of heart rate rhythms influenced by ANS and PNS interactions. Standard heart rate variability (HRV) frequency domain analysis allows the division of the electrocardiographic (ECG) signal into clinically relevant frequency components associated with the autonomic nervous system. Secondly, studies in music to date have examined different pre-performance periods, ranging anywhere from 100 seconds to 5 minutes beforehand, and have monitored HR at different intervals, which impacts the capacity to detect any subtle changes in the mechanisms governing cardiovascular response. According to the recommendations of the Task Force of the European Society of Cardiology (Malik et al., 1996), recordings of the ECG for HRV analysis should be standardized to a minimum of 5-minute epochs to achieve a suitable window in which to see short-term fluctuations and to assist in consistent assessment across studies. For these reasons, it is important to address possibilities of collecting, analyzing, and interpreting cardiovascular function in greater detail.

Analysing Physiological Data in Musical Performance

HRV is usually monitored through the ECG, a reasonably non-invasive method that detects the regularity of beat-to-beat (R-R) or normal-to-normal (N-N) intervals (Berntson & Cacciappo, 2004). ECG captures the atrial depolarization (P wave), the depolarization of the right and left ventricles (QRS complex), and the recovery of the ventricles (T wave), with another P wave completing the cycle. The index of interest in stress detection is the R-R interval: the time difference between one ventricular depolarization to another. ECGs can be obtained using electrodes placed on the skin's surface that either are physically connected to an external recording device or connected by a wireless belt equipped with sensors and strapped around designated body locations. The signal is then transformed from a continuous signal into a discrete one (Malik et al., 1996). The accuracy of the signal depends on the recorded sampling frequency (known as the number of samples per unit of time), and the adequate preparation of the equipment to monitor the user. Low sampling frequencies (100Hz) lack a complete reconstruction of the recorded signal (Kamath et al., 2013), while other factors can adulterate the quality of the data, such as excessive body movement, abnormal breathing, muscle contraction, or the interference between the participant and the recording device. Such additional noise may produce artifacts or ectopic beats. The former are beats occurring outside of the patterns of the R-R fluctuations; the latter are extra beats that results in shortened R-R intervals, a compensational delay and an extended R-R interval, leading to a minimized stroke volume that affects at least 10-30 beats (Urbanowicz et al., 2007). Both impact the data and can mislead results if not spotted before further analysis (Nayak et al., 2012).

Once the data have been cleaned where necessary, they are ready to be processed. The ECG can be examined in terms of (1) Time domain analysis (Billman, 2011), (2) Frequency domain analysis (Kamath et al., 2013), and (3) Sample entropy measures drawn from complexity science (Costa et al., 2002, 2005).

Time Domain Analysis

Time domain analysis examines physiological signals and their progression with respect to time. Among others, it determines the average R-R (or N-N) intervals (AVNN), their standard deviation (SDNN), the root mean square of successive differences of R-R intervals (RMSSD), and the (percentage of) difference between adjacent R-R intervals that are greater than 50ms ([p]NN50) (Malik et al., 1996). Each parameter gives insights into the degree of the sympathovagal balance (see Table 1). In a relaxed state, for instance, the AVNN ranges between 0.6 and 0.8 seconds, with small SDNNs and RMSSD and a larger percentage of NN50. Although Time domain analysis does not require a high degree of 'stationarity' of the data, and fails to investigate R-R intervals based on their frequency distribution, it provides an overall marker of the HRV.

Table 1: Features extracted from Time domain analysis (Stein et al., 1994; Kleiger et al., 2005; Malik et al., 1996)

Abbreviation	Definition	Function
AVNN (ms)	Average of N-N intervals for period interest	Can convert to average HR of N-N intervals (HR=60,000/ AVNN)
SDNN (ms)	Standard deviation of N-N intervals for period of interest	Reflects total HRV
SDANN (ms)	Standard deviation of AVNN for 5-minute intervals for period of interest	Reflects primarily circadian HRV
SDNNIDX (ms)	Average of 5-minute standard deviations of N-N intervals for period of interest	Reflects average short-term HRV and combined SNS and PNS influences
pNN50 (%)	Percentage of N intervals >50ms different from previous (N-N) for period of interest	With normal sinus rhythm reflects vagal activity
RMSSD (ms)	Root mean square of successive differences of N-N intervals for period of interest	With normal sinus rhythm reflects vagal activity

Frequency Domain Analysis

The Frequency domain analysis investigates R-R fluctuations (i.e., HRV) in terms of their frequency spectrum. These are detected from the power spectral density (PSD), a measure that represents signals as frequency distributions. Established frequencies components consist of low frequencies (LF) and high frequencies (HF). LFs are those observed between 0.04-0.15 Hertz (Hz) and are associated with the SNS and baroreceptor reflex (i.e., blood pressure). HFs (0.15-0.4 Hz) are believed to provide information of the vagal modulation (i.e., PNS) and respiratory sinus arrhythmia (alterations in HRV due to naturally occurring breathing) (Malik et al., 1996). The ratio between the LF and HF power, the LF/HF ratio, indicates the global sympathovagal balance.

Other less investigated frequencies are the ultra low (ranging from $1.15x10^{-5}$ to 0.0033Hz) and very low frequencies (0.0033 to 0.04 Hz). Ultra low frequencies are assumed to reflect neuroendocrine rhythms and are obtained in long-term recordings (e.g., over 24 hours). Very low frequencies are believed to define the renin-angiotensin system (i.e., blood pressure regulation) and the chemoreceptor and thermoregulatory activity. Within shorter recordings (5-10 minutes), it has also been suggested to represent negative emotions, such as worry and rumination (Berntson et al., 1997; Bernston & Cacioppo, 2004; Bilge et al., 1999; Cacioppo et al., 1994; Taylor et al., 1998) (see Table 2).

The most commonly applied methods to compute PSDs are the Fast Fourier transformation (FFT) and the autoregressive model estimation (AR) (Malliani, 1998; Malliani et al., 1991). The FFT usually assumes an *a priori* understanding of frequency bands of interest. The AR is more complex to carry out, yet "[...] decompose the overall spectrum into single spectral components [...] providing automatically the central frequency and associated power without the need for *a priori* assumptions" (Montano et al., 2009, p. 73). The method used depends on the design of the study. Frequency domain analysis requires 'stationary' data, and ideally an infinite number of samples. HRV, however, is only collectable in blocks (or time windows) and contains outliers (or non-harmonics) produced by artifacts or ectopic beats. This causes a 'frequency leakage,' a misinterpretation of the data due to limited time windows and outliers. One way to minimize frequency leakage is to filter the data, and the most commonly used filter methods are the Butterworth filter, the cubic spline interpolation, and the Median filter. The Butterworth filter is defined by two parameters: the cut-off frequency and the order. While the cut-off frequency delimits the boundary of the frequency spectrum, the order determines the level of sharpness from pass band to stop band. Pass and stop bands are represented by a range of frequencies, being either let through or rejected, and when combined with the cut-off frequency, produce a monotonic filter response (Robertson & Dowling, 2003). The cubic spline interpolation divides a signal into small subsections and connects approximations between oscillations (i.e., piecewise polynomial interpolation), achieving an overall smoothness (Haghighi & Roohi, 2012). The median filter combines median values of the signal, making the data less vulnerable to outliers. All three methods enhance the data quality and help to facilitate further processing.

Table 2: Features extracted from Frequency domain analysis (Stein et al., 1994; Malik et al., 1996)

Abbreviation	**Explanation**	**Function**
TP	Total power over measured period	Reflects total HRV
ULF	Ultra low frequency power measures rhythms greater than 5 minutes	Reflects circadian HRV
VLF	Very low frequency power measures rhythms between every 25 seconds and every 5 minutes, (i.e. 0.0033-0.04 Hz)	Reflects vagal and renin-angiotensin system effects on HR
LF	LF measures HR rhythms from 2.5 to 9 cycles/minute (i.e. 0.04-0.15Hz)	Reflect combination of SNS and PNS influences. Captures baroreflex rhythms
HF	HF captures variations in HR due to respiratory sinus arrhythmia at 9-14 cycles/minute (i.e. 015-0.4Hz). Averaged over 5-minute periods or less	Under normal circumstances reflects vagal activity
LF/HF	LF/HF average over 5-minute periods or less	Purpose to reflect SNS/PNS balance
LFnu	[LF/(TP-VLF)] for the measured period (5-minutes or less). Some calculate LF/(LF+HF)	Purpose to reflect the SNS activity
HFnu	[HF/(TP-VLF)] for the measured period (5-minutes or less). Some calculate HF/(LF+HF)	Purpose to reflect the PNS activity

Sample Entropy: Moving Beyond Standard Analysis Methods

Both Time and Frequency domain methods have their limitations: Time domain analysis only offers a preliminary insight into HRV; features extracted from the frequency domain analysis represents changes in the sympathetic and parasympathetic nervous system. While the LFs in the HRV are believed to show increased power during physically and mentally stressful tasks, a pronounced activity has also been detected during more relaxed states, such as sleep or deep breathing (Rahman et al., 2011). Moreover, some studies have observed an opposite effect: decreased LF power during taxing demands (Billman, 2013). A second constraint is the discharge of the impact of breathing. There is a coupling effect between the (complexity of) heart rate and breathing (Bernardi et al., 2001; Schulz et al., 2015; Tharion et al., 2012), which should ideally be evaluated by providing a respiration-adjusted index of the HRV (Rahman et al., 2011; Widjaja et al., 2014).

An alternative approach to analyse HRV is available through the field of complexity science (Costa et al., 2002, 2005). Complexity science assumes that organic systems are unpredictable, complex, and infinitely diverse. They can be unexpected and disorderly, driven by negative, but also positive environmental feedback. Complexity science considers these aspects by investigating biosignals in terms of their 'linearity,' 'disorder' (i.e., complexity), and 'magnitude' (i.e., scale). This is achieved by the calculation of the signal's multiscale sample entropy (MSE), a method that offers a sophisticated statistic to expand the knowledge of cardiovascular functioning. To put this into context, low MSE refers to an extremely regular physiological time series, and has been shown to appear alongside cardiovascular diseases (Shin et al., 2006; Tuzcu et al., 2006) in aging individuals (Takahashi et al., 2006), and under conditions of musical performance stress (Williamon et al., 2013).

It is important to understand how to measure and interpret the psychophysiological experiences of musicians accurately, prior to and during performance. For the musician, however, interventions to assist those that experience debilitating stress levels need to be highlighted. This will help them find coping strategies to facilitate performing to their potential. While many coping strategies exist, the next section considers the specific role of physical activity in the reduction of elevated psychophysiological states based on the available research.

Physical Interventions Designed to Optimize Performance Outcomes

Many musicians experience performances with apprehension and increased physical and mental stress, such as elevated heart rate or intense feelings of anxiety (Kenny, 2011). In order to improve musicians' performance outcomes and help them manage the challenges of performing, a wide variety of strategies and techniques are now being applied, including cognitive-behavioral interventions and mental skills training, but also more applied strategies, such as physical exercise.

In particular, physical training has been shown to benefit cognitive functioning (Vancampfort et al., 2011; Carek et al., 2011; Ströhle, 2009), cardiovascular functioning (Blumenthal et al., 2005; Hambrecht et al., 2003), and stress management (Brownley et al., 2003; Chafin et al., 2008). Defined as "[...] participation in a program of regular, structured, physical exertion of varying degrees of intensity designed to increase heart rate and/or muscle strength" (Edenfield & Blumenthal, 2011, p. 300; Lippincott & Wilkis, 2014), two types of exercise exist: aerobic and anaerobic. Aerobic exercise implies activities used with oxygen, such as brisk walking or jogging, supplying the cells with energy. Anaerobic exercise triggers lactic acid formation, a process by which enhanced metabolism of muscle glycogen occurs, building up muscle mass. This is produced by a high intensity workout, such as weight lifting or sprinting. Aerobic exercise is especially believed to facilitate a cross-stressor adaption (i.e., Cross-Stressor Adaptation hypothesis; [Sothmann et al., 1996]), an increased tolerance towards familiar and novel physiological and psychological stressors. These adaptions are also assumed to reduce the sympathetic nervous activity and increase the parasympathetic nervous activity.

Indeed, stress research overall confirms the positive effect of exercise on physiological and psychological responses to stress; yet, it is still debated which type, frequency, and duration of physical interventions optimize physical and mental responses to (performance) stress. Petruzzello et al. (1991), for instance, examined in a meta-review the relationship between various types of anaerobic and aerobic exercises (e.g., weight-training versus walking, swimming) and state and trait anxiety inventories (e.g., Spielberger et al., 1983). He reviewed 104 studies, considering the categories: (i) exercise (aerobic vs. non-aerobic, duration and intensity of the exercise carried out), (ii) design of the study (pre-post-/multiple-assessment, control versus comparison group (e.g., meditation, relaxation), and (iii) type of participant (age and mental health status). Based on 408 effect sizes and a sample size of 3048 participants, the results showed a significant overall relationship between exercise and state anxiety reduction from pre-test to post-test interventions after aerobic but not anaerobic exercise. Both acute and chronic exercise was equally effective, yet, compared with other interventions, such as yoga, the effect diminished. Higher exercise intensity (measured in $\%HR_{max}$ or $\%VO_{2max}$) caused greater reduction in state anxiety, while the time the state anxiety inventory was administered (< 5 minutes versus > 20 minutes) did not have an impact. In terms of exercise and trait anxiety, a significant overall response between exercise and anxiety reduction was observed, with again, a significant effect for aerobic, but not anaerobic exercise. This outcome remained stable when other interventions were considered. Higher exercise intensity (measured in $\%HR_{max}$ or $\%VO_{2max}$) carried out for at least 20 minutes produced greater changes in anxiety reduction, especially in participants: (i) suffering from high levels of anxiety; and (ii) between 31-45 years.

Long et al.'s (1995) meta-review investigated the effect of exercise carried out at least 3 times per week, 20 minutes in duration, over a period of 6 weeks. Exclusion criteria were participants under 18 years and non-standardized physiological and psychological measures. Inclusion criteria were: (i) experimental or quasi-experimental studies; (ii) within versus within-between subject designs (pre-test versus post-test; pre-test versus post-test versus experimental versus control group); and (iii) self-reports on state and trait anxiety that fulfil psychometric criteria (e.g., reliability and validity). 26 studies investigating pre-test versus post-test effects and 20 studies comparing exercise with other interventions, such as yoga, were selected. The pre-test and post-test comparisons revealed: (i) a significant overall effect between exercise and anxiety reduction; (ii) a greater impact for higher than lower anxious participants; and (iii) a greater effect for men than for women. The effect of exercise on state and trait anxiety outcome was similar, with shorter interventions (6-8 weeks) producing a greater impact than long-term interventions (> 12 weeks). Weekly sessions performed less than 3 times were more effective than those carried out more than 3 times. In terms of exercise during contrast group assessments (between groups with pre-test and post-test assessment), results showed: (i) a significant overall effect between exercise and anxiety reduction; (ii) a greater impact for higher than lower anxious participants; and (iii) greater results for men than for women. Shorter intervention (6-8 weeks) produced a greater anxiety reduction than long-term interventions (> 12 weeks), and weekly sessions performed more than 3 times were more effective than those carried out less than 3 times.

Lastly, Conn (2010) investigated the relation between exercise (e.g., supervised [educational or motivational] exercise sessions) and anxiety reduction among healthy participants on a long-term basis. He selected 19 studies published between 1983 and 2008. Exclusion criteria were: (i) participants with mental disorders; and (ii) studies that did not measure the long-term effect of exercise on anxiety. Studies were coded with regards to sample size, participants' characteristics, methods, and type of interventions applied. Results demonstrated significant effects between exercise and anxiety reduction, while the time where the self-reports were collected did not significantly change the anxiety outcome (< 90 days); however, group intervention was significantly less effective than exercise on a one-to-one basis, as well as when exercise was carried out in lower rather than higher intensity. The impact of exercise on anxiety reduction was reinforced by supervised training compared with unsupervised training, as well as when carried out in a gym rather than at home.

Based on these reviews, it can be concluded that exercise causes a noticeable, yet heterogeneous effect on anxiety. Such discrepancies were linked with variation in: (i) type, lengths, and frequency of exercise; (ii) exercise compared with other interventions; (iii) age; and (iv) gender, with older and male participants experiencing greater benefit than their younger and female counterparts.

Surprisingly, not many studies have evaluated the effect of physical exercise on musical performance stress, even though its application is straightforward and relatively easy to apply. Wasley et al. (2011) assessed 46 musicians on two days, one week apart. During the first unit, height, weight, baseline heart rate, blood pressure, state anxiety, as well as musicians' perceived physical activity, were taken. In addition, a cycling test at 60, 90, and 120 Watt was conducted. The index of physical fitness was expressed in VO_{2max}, known as the maximum rate of oxygen consumption during gradually increasing exercise. In the second unit, musicians performed in front of a jury with the possibility to win a prize based on the performance quality. Scheduled 20 minutes prior to the performance, musicians' blood pressure was taken 15, 10, and 5 minutes before the performance, while ECG was monitored throughout. The subjective feeling of anxiety was assessed before and after. Results showed that musicians ranged 'normally' on the BMI index and did approximately 210 minutes of moderate to vigorous sport per week. Anxiety, heart rate, and blood pressure were significantly elevated during the pre-performance period in the high-stress condition, though no specification of the outcome between the three measure point (15, 10, and 5 minutes) was provided. The degree of fitness was positively correlated with heart rate (variability), while a negative correlation between fitness level and state anxiety occurred. Fitter musicians exhibited a more pronounced decrease in state anxiety during the post-performance period, suggesting an association between fitness and psychological recovery. Future studies may follow up on Wasley's work, by also comparing participants fitness levels, and through experimental manipulation of musicians' degree of physical activity.

For musicians, it is important that they have excellent command over physical and mental skills in order to perform. They need to be in a sufficient physiological and psychological health-state, which is, ideally, trained from an early age and preceded by a screening that includes a rigorous assessment of physical fitness, coping abilities, and the tendency toward health promoting responsibilities. While schools of music and conservatoires have now begun to accept the need for such tailored intervention programs (Clark et al., 2013), more research is needed to help individual musicians manage stress and optimize their performance outcomes.

Acknowledgement

The work reported in this chapter was supported in part by the Peter Sowerby Foundation.

References

Abel, J. L., & Larkin, K. T. (1990). Anticipation of performance among musicians: physiological arousal, confidence, and state-anxiety. *Psychology of Music, 18*, 171-82.

Appel, S. S. (1976). Modifying solo performance anxiety in adult pianists. *Journal of Music Therapy, 13*, 2.

Bernardi, L., Porta, A., Gabutti, A., Spicuzza, L., & Sleight, P. (2001). Modulartory effects of respiration. *Autonomic Neuroscience: Basic and Clinical, 90*, 47-56.

Berntson, G. G., Bigger, J. T., Jr., Eckberg, D. L., Grossman, P., Kaufmann, P. G., Malik, M., Nagaraja, H. N., Porges, S. W., Saul, J. P., Stone, P. H., &Van der Molen, M. W. (1997). Heart rate variability: origins, methods, and interpretive caveats. *Psychophysiology, 34*, 623-48.

Berntson, G. G., & Cacioppo, J. T. (2004). Heart rate variability: stress and psychiatric conditions. In M. Malik & J. Camm (Eds.), *Dynamic Electrocardiograph*. Hoboken: Wiley-Blackwell.

Bilge, A. R., Stein, P. K., Domitrovich, P. P., Gérard, P. L., Rottman, J. N., Kleiger, R. E., Kulbertus, H. E., & Piérard, L. A. (1999). Assessment of ultra low frequency band power of heart rate variability: validation of alternative methods. *International Journal of Cardiology, 71*, 1-6.

Billman, G. E. (2011). Heart rate variability - a historical perspective. *Frontiers in Physiology, 2*: 86.

Billman, G. E. (2013). The LF/HF ratio does not accurately measure cardiac sympatho-vagal balance. *Frontiers in Physiology, 4*: 26.

Blumenthal, J. A., Sherwood, A., Babyak, M. A., Watkins, L. L., Waugh, R., Georgiades, A., Bacon, S. L., Hayano, J., Coleman, R. E., & Hinderliter, A. (2005). Effects of exercise and stress management training on markers of cardiovascular risk in patients with ischemic heart disease. *Journal of the American Medical Association, 293*, 1626-34.

Brotons, M. (1994). Effects of performing conditions on music performance anxiety and performance quality. *Journal of Music Therapy, 31*, 63-81.

Brownley, K. A., Hinderliter, A. L., West, S. G., Girdler, S. S., Sherwood, A., & Light, K. C. (2003). Sympathoadrenergic mechanisms in reduced hemodynamic stress responses after exercise. *Medicine and Science in Sports and Exercise, 35*, 978-86.

Cacioppo, J. T., Uchino, B. N., & Berntson, G. G. (1994). Individual differences in the autonomic origins of heart rate reactivity: the psychometrics of respiratory sinues arrythmia and preejection period. *Psychophysiology, 31*, 412-419.

Carek, P. J., Laibstain, S. E., & Carek, S. M. (2011). Exercise for the treatment of depression and anxiety. *The International Journal of Psychiatry in Medicine, 41*(1), 15-28.

Cella, D. F., & Perry, S. W. (1986). Reliability and concurrent validity of three visual-analog mood scales. *Psychological Reports, 59*, 827-33.

Chafin, S., Christenfeld, N., & Gerin, W. (2008). Improving cardiovascular recovery from stress with bried poststress exercise. *Health Psychology, 27*, 64-72.

Clark, T., Williamon, A., & Redding, E. (2013). The value of health screening in music schools and conservatoires. *Clinical Rheumatology, 32*, 497-500.

Conn, V. S. (2010). Anxiety outcomes after physical activity interventions: Meta-analysis findings. *Nursing Research, 59*, 224-31.

Costa, M., Goldberger, A., & Peng, C. K. (2002). Multiscale entropy analysis of complex physiologic time series. *Physical Review Letters, 89.*
Costa, M., Goldberger, A., & Peng, C. K. (2005). Multiscale entropy analysis of biological signals. *Physical Review E, 71.*
Craske, M. G., & Craig, K. D. (1984). Musical performance anxiety: the three-systems model and self-efficacy theory. *Behaviour Research and Therapy, 22*, 267-80.
Edenfield, T. M., & Blumenthal, J. A. (2011). Exercise and stress reduction. In R. J. Contrada & A. Baum (Eds.), *The Handbook of Stress Science: Biology, Psychology, and Health.* New York: Springer Publishing Company.
Fredrikson, M., & Gunnarsson, R. (1992). Psychobiology of stage fright: the effect of public performance on neuroendocrine, cardiovascular and subjective reactions. *Biological Psychology, 33*, 51-61.
Haghighi, A. R., & Roohi, M. (2012). The fractional cubic spline interpolation without using the derivate values. *Indian Journal of Science and Technology, 5*(10), 3433-3439.
Hambrecht, R., Adams, V., Erbs, S., Linke, A., Krankel, N., Shu, Y., Baither, Y., Gielen, S., Thiele, H., Gummert, J. F., Mohr, F. W., & Schuler, G. (2003). Regular physical activity improves endothelial function in patients with coronary artery disease by increasing phosphorylation of endothelial nitric oxide synthase. *Circulation, 107*, 3152-8.
Kamath, M. V., Watanabe, M. A., & Upton, A. (2013). *Heart Rate Variability (HRV) Signal Analysis: Clinical Applications.* Boca Raton: Taylor & Francis.
Kenny, D. (2011). *The Psychology of Music and Performance Anxiety.* Oxford: Oxford University Press.
Kleiger, R. E., Stein, P. K., & Bigger, J. T., Jr. (2005). Heart rate variability: measurement and clinical utility. *Annals of Noninvasive Electrocardiology, 10*(1), 88-101.
Lazarus, R. S., & Folkman, S. (1984). *Stress, Appraisal, and Coping.* New York: Springer.
Lippincott, W. K., & Wilkis, W. (2014). *ACM's Guidelines for exercise testing and prescription.* Baltimore: MD.
Long, B. C., & van Stavel, R. (1995). Effects of exercise training on anxiety: a meta-analysis. *Journal of Applied Sport Psychology, 7*(2), 167-189.
Malik, M., Bigger, J. T., Camm, J. A., Kleiger, R. E., Malliani, A., Moss, J. A., & Schwartz, P. J. (1996). Heart rate variability: standards of measurement, physiological interpretation, and clinical use. Task force of the European Society of Cardiology & The North American Society of Pacing and Electrophysiology. *Circulation, 93*, 1043-65.
Malliani, A. (1999). The pattern of sympathovagal balance explored in the frequency domain. *News in Physiological Sciences, 14*, 111-17.
Malliani, A., Pagani, M., Lombardi, F., & Cerutti, S. (1991). Cardiovascular neural regulation explored in the frequency domain. *Circulation, 84*, 482-92.
Montano, N., Porta, A., Cogliati, C., Costantino, G., Tobaldini, E., Casali, K. R., & Iellamo, F. (2009). Heart rate variability explored in the frequency domain: a tool to investigate the link between heart and behavior. *Neuroscience & Biobehavioral Reviews, 33*, 71-80.
Nayak, S., Soni, M. K., & Bansal, D. (2012). Filtering techniques for ECG signal processing. *International Journal of Research in Engineering and Applied Sciences, 2*, 671-79.

Petruzzello, S. J., Landers, D. M., Hatfiled, B. D., Kubitz, K. A., & Salazar, W. (1991). A meta-analysis on the anxety-reducing effects of acute and chronic exercise. Outcomes and mechanisms. *Sports Medicine, 3*, 143-82.

Rahman, F., Pechnik, S., Gross, D., Sewell, L., & Goldstein, D. S. (2011). Low frequency power of heart rate variability reflects baroreflex function, not cardiac sympathetic innervation. *Clinical Autonomic Research, 21*, 133-41.

Robertson, D. G., & Dowling, J. J. (2003). Design and responses of Butterworth and critically damped digital filters. *Journal of Electromyography and Kinesiology, 13*(6), 569-573.

Sarafino, E. P. (1996). *Health Psychology: Biopsychosocial Interactions.* John Wiley & Sons: Hoboken.

Schulz, S., Bär, K.-J., & Voss, A. (2015). Analyses of heart rate, respiration and cardiorespiratory coupling in patients with schizophrenia. *Entropy, 17*, 483-501.

Shin, D.-G., Yoo, C.-S., Yi, S.-H., Bae, J.-H., Kim, Y.-J., Park, J.-S., & Hong, G.-R. (2006). Prediction of paroxysmal atrial fibrillation using nonlinear analysis of the R-R interval dynamics before the spontaneous onset of atrial fibrillation. *Circulation Journal, 70*, 94-99.

Sothmann, M. S., Buckworth, J., Claytor, R. P., Cox, R. H., White-Welkley, J. E., & Dishman, R. K. (1996). Exercise training and the cross-stressor adaption hypothesis. *Exercise & Sport Sciences Reviews, 24*, 267-88.

Spielberger, C. D., Gorsuch, R. L., Lushene, R., Vagg, P. R., & Jacobs, G. A. (1983). *Manual for the State-Trait Anxiety Inventory.* Palo Alto, CA: Consulting Psychologists Press.

Stein, P. K., Bosner, M. S., Kleiger, R. E., & Conger, B. M. (1994). Heart rate variability: A measure of cardiac autonomic tone. *American Heart Journal, 127*, 1376-81.

Ströhle, A. (2009). Physical activity, exercise, depression and anxiety disorders. *Journal of Neural Transmission, 116*, 777-84.

Takahashi, T., Cho, R. Y., Murata, T., Mizuno, T., Kikuchi, M., Mizukami, K., et al. (2009). Age-related variation in EEG complexity to photic stimulation: a multiscale entropy analysis. *Clinical Neurophysiology, 120*(3), 476-483.

Taylor, J. A., Carr, D. L., Myers, C. W., & Eckberg, D. L. (1998). Mechanisms underlying very-low-frequency RR-interval oscillations in humans. *Circulation, 98*, 547-55.

Tharion, E., Samuel, P., Rajalakshmi, R., Gnanasenthil, G., & Subramanian, R. K. (2012). Influence of deep breathing exercise on spontaneous respiratory rate and heart rate variability: A randomised controlled trial in healthy subjects. *Indian Journal of Physiology and Pharmacology, 56*, 80-87.

Tuzcu, V., Nas, S., Borklu, T., & Ugur, A. (2006). Decrease in the heart rate complexity prior to the onset of atrial fibrillation. *Europace, 8*, 398-402.

Urbanowicz, K., Żebrowski, J. J., Baranowski, R., & Hołyst, J. A. (2007). How random is your heart beat? *Physica A: Statistical Mechanics and its Applications, 384*, 439-47.

Vancampfort, D., De Hert, M., Knapen, J., Wampers, M., Demunter, H., Deckx, S., Maurissen, K., & Probst, M. (2011). State anxiety, psychological stress and positive well-being responses to yoga and aerobic exercise in people with schizophrenia: a pilot study. *Disability and Rehabilitation, 33*, 684-9.

Wasley, D., Taylor, A., Backx, A., & Williamon, A. (2011). Influence of fitness and physical activity on cardiovascular reactivity to musical performance. *A Journal of Prevention, Assessment and Rehabilitation, 41*, 27-32.

Widjaja, D., Caicedo, A., Vlemincx, E., Van Diest, I., & Van Huffel, S. (2014). Separation of respiratory influences from the tachogram: a methodological evaluation. *Plos One, 9*, e101713.
Williamon, A., Aufegger, L., Wasley, D., Looney, D., & Mandic, D. P. (2013). Complexity of physiological responses decreases in high-stress musical performance. *Journal of the Royal Society Interface, 10*, 20130719.
Wolpe, J. (1969). *The Practice of Behavior Therapy.* New York: Pergamon Press.
Yoshie, M., Kudo, K., Murakoshi, T., & Ohtsuki, T. (2009). Music performance anxiety in skilled pianists: effects of social-evaluative performance situation on subjective, autonomic, and electromyographic reactions. *Experimental Brain Research, 199*, 117-26.

Psychological Skills to Support Performance Under Pressure

The need to perform at one's best in high pressure situations may for many performers, at some point in their learning and professional careers, trigger debilitating performance anxiety. This is associated with marked emotional distress, impaired performance and performance avoidance, presenting a substantial barrier to learning and occupational health. Yet, our understanding of valid and reliable interventions to reduce performance anxiety for adolescents and young adults who are training for a performance career is minimal. This chapter summarizes findings from performance psychology skills training programs which translate the psychology of optimum performance in elite sports to the musicians' training process. By enhancing performers' self-knowledge, combined with the application of anxiety management strategies when under pressure, these studies show that the negative effects of performance anxiety can be reduced. Commensurate with this, performance resilience increases through improved confidence, self-belief, courage, persistence and ability to recover from mistakes.

Performing music at a professional level is one of the most complex human performance tasks, incorporating both cognitive (such as mathematical problem solving), and sensorimotor skills (such as sport performance) (Münte, Altenmüller, & Jäncke, 2002). A concert pianist, for example, may bimanually coordinate the production of up to 1,800 notes per minute. The need for performance excellence exposes musicians to significant physical and psychological stress and injury throughout their career. Performance anxiety is the most significant psychological issue for performing musicians of any age (Kenny, 2011), with up to 75% of students experiencing some form of performance anxiety (Britsch, 2005). This can be a significant blow to a performer's psychological well-being and efforts to achieve optimal performance (Fehm & Schmidt, 2006).

A recent population study of Australian musicians found performance anxiety to be the most common risk to mental health, particularly for females and those younger than 30 years of age (Kenny, Driscoll, & Ackermann, 2014). Two key triggers were cited as pressure from self (suggesting maladaptive perfectionism), and worries about the negative impact of excessive physiological arousal prior to or during a performance.

In contrast, performers with a strong sense of competence and control over their physical and psychological well-being and performance skills enhance their capacity to exceed their average level of performance and perform well (Kenny, 2011; Williamon, 2004).

Anxiety and Performance Quality

It was originally believed that a simple relationship existed between anxiety and performance, namely that anxiety impairs the quality of performance, summarized by the "inverted U" law of arousal and performance (Hancock & Ganey, 2003). In a reinterpretation of the 1908 "Yerkes-Dodson Law" or inverted-U representation of the relationship of arousal to performance, Mornell (2016) reframed the original narrow conceptualization which included physiological "arousal" as "physiological and emotional energy, concentration and control," capturing the more involved processes underlying the psychological aspects of performance.

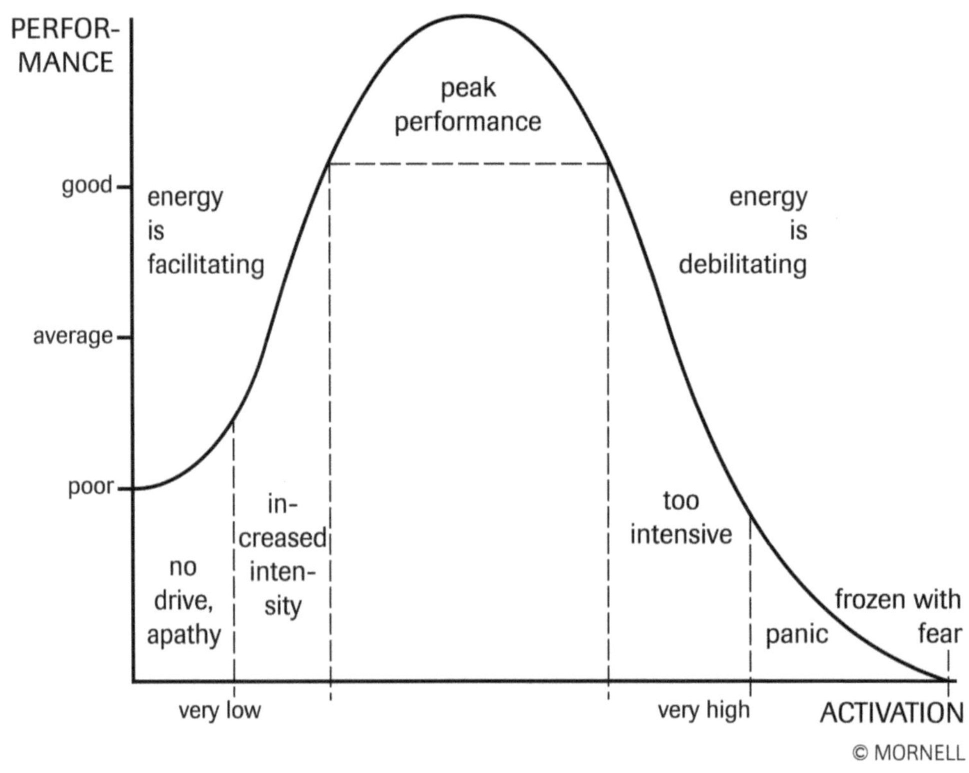

Figure 1: Inverted-U relationship between performance energy activation and performance outcome (see: Mornell, 2002)

When performing at their best (known as a peak performance), performers report feeling highly confident, without fear of failure. They are in control of their emotions, thoughts and physiological arousal, and are highly energized yet physically and mentally relaxed. They are totally immersed in the activity and execute tasks effortlessly and automatically. Successful performers view anxiety as beneficial by reframing it as performance energy or excitement (Brooks, 2014). Poor performance is associated with self-doubt, feeling over- or under-aroused, loss of concentration, and difficulty focusing on the outcome of the performance (Krane & Williams, 2010). In high stress states, individuals have greater difficulty learning, remembering, and executing skilled behavior (Kim & Diamond, 2002).

The parsimony of the inverted-U model has no doubt encouraged its continued use in performance coaching to summarize the facilitating and debilitating effects of anxiety on performance. Yet, equivocal results of research aiming to attenuate performance anxiety through psychological intervention attest to a more complex, extended stress-performance relationship. In some early studies, musicians' performance anxiety decreased and performance quality improved following psychotherapeutic intervention (Appel, 1976; Roland, 1994; Sweeney & Horan, 1982). In contrast, other studies have found no improvements in performance quality following behavioral interventions (Deen, 2000; Mansberger, 1988; Reitman, 2001; Wardle, 1975). For example, whilst Osborne and colleagues (Braden, Osborne, & Wilson, 2015; Osborne, 2013; Osborne, Kenny, & Cooksey, 2007) found significant improvement in self-reported MPA (Music Performance Anxiety) following cognitive-behavioral interventions in adolescent musicians, there appeared to be no effect on performance quality as rated by external judges.

It is now understood that a moderate amount of anxiety enhances performance when an individual's skill level matches the performance demands of the situation (Jackson & Csikszentmihalyi, 1999) and the individual interprets that anxiety positively (Jones, Swain, & Hardy, 1993). For example, Hancock and Warm's (1989) dynamic model of stress and sustained attention provides an "extended-U" conceptualization by incorporating both physiological and psychological adaptability to account for the complexity of stress on human performance capability (Hancock & Ganey, 2003). Stress-performance conceptualizations such as these attribute good, optimal performance to psychological processes which extend beyond technical competency.

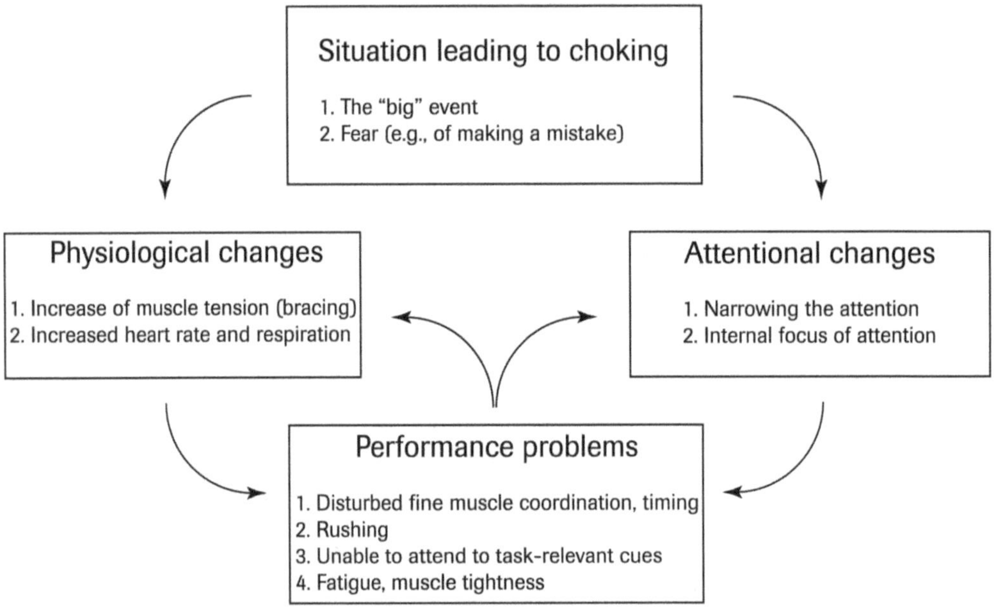

Figure 2: Stress-performance model of choking. Adapted from Nideffer and Sagal (2001, p. 317).

"Choking under pressure" describes performance decrements under pressure conditions despite the individual striving to perform well (Beckmann, Gröpel, & Ehrlenspiel, 2013). Pre-performance routines have a significant bearing on the likelihood of choking under pressure in sport (Hanton, Thomas, & Maynard, 2004). In sports with a preparatory period, such as golf, consistency and reproducibility of a pre-shot routine is one of the most important differences that distinguish experts from novices (Milton, Solodkin, Hlustik, & Small, 2007). Pre-performance routines which combine psychological and behavioral components decrease the likelihood of choking, compared to singular methods of deep breathing, cue word, temporal duration or no routine control (Mesagno & Mullane-Grant, 2010). In the previous *Art in Motion II*, Beckmann et.al., (2013) outlined their research showing that pre-performance routines which increase right-hemisphere activation through hemispheric-specific priming attenuated motor skill failure under pressure conditions in gymnasts. Stronger left-hemispheric inhibition is associated with better performance in golf-putting and other automatic behaviors such as music performance. This is because increased pressure engenders the desire to perform well, directing attention inward to conscious monitoring of the execution of motor behavior (Figure 2). Attention to execution at the step-by-step level (left-hemisphere dominant) interferes with the execution of automatic processes (right-hemisphere dominant), causing the expert performer to regress to the level of novice in early, cognitive phases of motor learning (Gucciardi & Dimmock, 2008).

Strategies for Success

Given that "the mind of expert motor performance is cool and focused" (Milton et al., 2007, p. 804), which is in direct opposition to the physiological and attentional changes experienced when performing under pressure, what can we do to achieve that mental state? In adult and adolescent musicians, the most effective anxiety treatment paradigm for MPA involves a combination of cognitive and behavioral techniques, including cognitive restructuring, attention control and behavioral rehearsal, relaxation and mental skills training (Osborne et al., 2007).

Performance psychology techniques drawn from the sports domain have a strong evidence base in performance enhancement in recent years following treatment with athletes for performance anxiety (Williams, 2010). Sport and music performance have many overlapping features: both require high levels of motor control and learning, mastery over mind and body, the necessity for implicit recall and smooth performance, as well as the presence of an audience, which for the performer may invoke enjoyment of excellence and/or psychological pressure (Yoshie, Kudo, Murakoshi, & Ohtsuki, 2009). Performance psychology techniques address the "cognitive, emotional, behavioral and psychophysiological inhibitors of consistent, excellent performance" to support individuals to "consistently perform in the upper range of their capabilities"(American Psychological Association, 2011, p. 9).

Music Performance Enhancement Program

To date, two studies have evaluated psychological skills programs for MPA in school settings. Osborne et al. (2007), implemented a combined group-individual cognitive-behavioral intervention program to reduce MPA and improve performance quality for 23 students at the Sydney Conservatorium of Music High School. Measures of MPA, state-trait anxiety, anxiety diagnoses, heart-rate and muscle tension were taken. Within one week students gave an unaccompanied solo performance of 2-3 minutes demonstrating their highest level of musical proficiency. Students were then randomly allocated to the Music Performance Enhancement Program (MPEP) as a cognitive-behavioral performance psychology intervention, or were required to perform the piece of music only as a behavior-exposure-only condition.

The MPEP was delivered over three 1-hour group sessions (weeks 1, 3, and 5) after school, and four 45 minute individual sessions (weeks 2, 4, 6, and 7) within school time. Group sessions covered topics which were followed up with students in their individual sessions: psychoeducation regarding the physiological, cognitive and behavioral aspects of MPA; relaxation strategies; goal setting; cognitive restructuring and self-talk; performance diaries; performance routines, choking, warm-up decrements, and paralysis-by-analysis. Within two weeks of completing the program all students completed the same self-report and diagnostic interview as pre-intervention, and performed the same piece of music as their first performance with accompanying physiological measurements.

Following the intervention, while there was no difference in performance quality, students who participated in the MPEP reported less MPA, performance fear, and avoidance than students who were in the behavior-exposure-only group. Notably, there were additional reductions in MPA for the adherent MPEP participants. The study also included a performance concert condition in which all students performed the same piece of music to an audience of friends, relatives, and Conservatorium staff one month after completing the program. Prior to this concert, state anxiety scores for the adherent group reduced dramatically, demonstrating ecological validity (Osborne, 2008).This study demonstrated that psychological skills training could significantly reduce MPA and performance avoidance within a school setting, with the additional benefit of reduced anxiety in other areas of students' lives.

Thinking Skills for Peak Performance

A subsequent study by Osborne and colleagues (Braden et al., 2015; Osborne, 2013), evaluated the performance psychology program *Thinking Skills for Peak Performance: Unleash your Potential!* (Brandon & Ivens, 2009) as an early intervention strategy for high school musicians.

This program builds students' self-knowledge in relation to achieving a peak performance. In this study, a randomized wait-list control design was used to delineate the effect of time from intervention which was not possible in the Osborne et. al., (2007) study. The eight-week program covered peak performance, personal strengths, planning and goal setting, motivation, positive self-talk, relaxation, imagery and visualization, stress regulation, and recovery after disappointments or setbacks. Three solo performances performed in front of two judges acted as both a behavioral exposure and outcome assessment of manifest MPA. Additional outcome measures included self-reported state (pre-performance) and trait (non-performance) MPA, motivation to learn and perform music, and expert ratings of performance quality and observable behavioral manifestations of MPA.

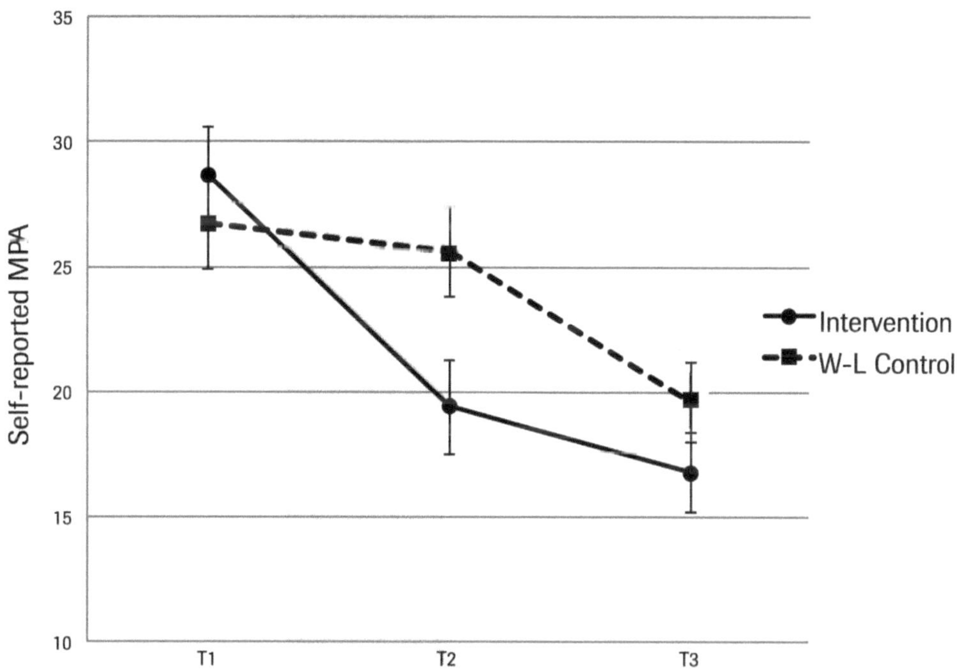

Figure 3: Self-rated MPA scores by group and time (mean scores with standard error bars, where higher scores indicate more anxiety). *Source:* From "Psychological intervention reduces self-reported performance anxiety in high school music students," by A. M. Braden, M. S. Osborne and S. J. Wilson, 2015, *Frontiers in Psychology, 6*, pp. 1-9. Reprinted with permission.

Self-reported MPA [using the *Music Performance Anxiety Inventory for Adolescents*, Osborne & Kenny, 2005] reduced significantly for all students after completing the program (see Figure 3 – Intervention group 1: T1-pre, T2-post, T3-follow-up; Control group 2: T2-pre, T3-post). MPA scores continued to decline two months after completing the program (T3) for the first intervention group, showing both maintenance of gains *and* continued improvement after program completion. There was some evidence of improved performance quality, yet the intervention had no effect on judge-rated behavioral manifestations of MPA.

Significant improvements were also observed in music performance resilience as measured by the Motivation and Engagement Scale for Music (MES-M, Martin, 2008). Improved self-efficacy scores indicated that students were more likely to want to perform in order to achieve a successful outcome, rather than to avoid failure. Performance preparation improved through increased planning, effort and persistence, showing improved autonomy in achieving successful outcomes, with a commensurate reduction in self-handicapping and withdrawal behaviors. In short, as a result of the intervention students' music performance preparation strategies became more self-regulated (Osborne, 2013).

Both studies (Braden et al., 2015; Osborne, 2013; Osborne et al., 2007) demonstrate that performance psychology programs which teach young musicians the mental skills needed to enhance personal confidence and competence to perform at one's best, lead to significant reductions in MPA when delivered within the school environment. Limited work has been undertaken in slightly older musicians in tertiary music performance education settings using performance psychology methodologies which frame anxiety positively, as a constructive energy for optimized performance. For this reason, Osborne and colleagues tested the efficacy of two different programs which translate the psychology of optimum performance in elite sports to musicians' higher education performance training. The point of difference in these studies compared to most studies in the literature with musicians in higher education performance training is that the underlying rationale of these two programs is *not* to actively *reduce* performance anxiety. Rather, they aim to achieve the highest level of performance through self-knowledge and the application strategies that trigger optimal performance behaviours *despite* anxiety.

Centering

Consistent with the dynamic model of stress and performance (Hancock & Ganey, 2003; Hancock & Warm, 1989), recent work has shown that a pre-performance routine which addresses the negative effects of performance anxiety focuses on two dominant drivers of MPA in musicians of all backgrounds, namely, negative, debilitating thoughts and excessive physiological arousal (Kenny 2011; Kenny et al., 2014). The aim of centering is to gain control over the destructive escalation in nervous energy (i.e., physiological arousal) and distracted attention toward external (e.g., audience, performance environment) and internal (e.g., anxious rumination) non-performance cues that can occur before and during high-pressure performance situations.

There are four main parts to this technique. First, identifying a clear intention of what is to be accomplished in the task, e.g., "I'm going to learn how to center," or "I'm going to nail the high D in the aria." Second, distributing one's body weight around the center of mass (approximately seven centimeters below and behind the navel), counteracts the natural tendency for physical tension to rise with increasing nerves and communicates a readiness to perform. Third, a breathing technique that is initially focused on taking deep breaths into the abdomen rather than the chest. When breathing out, performers release tension in muscle groups common in anxiety (e.g., neck and shoulder tension), as well as key muscles required in playing their instrument. Fourth, using a working model of the left and right brain hemispheres, performers are directed to shift attention away from the left hemisphere. Responsible for a verbally dominant analysis and judgment of the performance task, this hemisphere is helpful in learning a piece or specific technique, but a major obstacle to achieving the mental quiet necessary for optimal performance when it is flooded with stress-induced negative thought patterns. Instead, performers are instructed to engage the right brain hemisphere where the performer feels, hears and pictures him/herself performing well. The focus on positive images of the self is ideal for optimal performance, and mandatory for peak performance. For performing artists, the critical process is developing the ability to deftly switch between left and right brain, according to changing performance task demands (Greene, 2002).

In early 2013, the author conducted a study testing the efficacy of Greene's centering technique, supplemented by his workbook of *11 Strategies for Audition and Performance Success* for musicians (Greene, 2012; Osborne, Greene, & Immel, 2014). Pre-post intervention analyses from thirty-one Melbourne Conservatorium of Music students measured on the Performance Skills Inventory (PSI) demonstrated substantial benefits of the program across key psychological performance skills:

1. Performance energy - an overall reduction in anxiety and improvement in ability to manage excessive physiological arousal;
2. Preparation - improved ability to learn and memorize music and build more effective practice habits;
3. Confidence - more positive self-talk and enhanced ability beliefs;
4. Courage - greater willingness to confront fears and take risks;
5. Focus - increased ability to be mindful and "in the present," pay no mind to distractions;
6. Concentration - greater ability to quiet one's mind under the pressure of performance; and
7. Resilience - ability to let go of mistakes and regain equilibrium during performance.

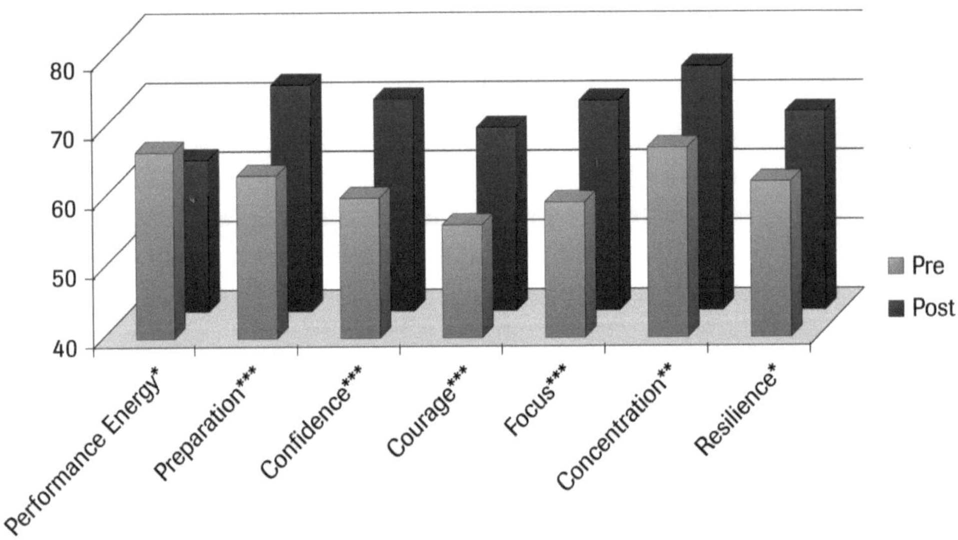

Figure 4: Pre- and post-program results for the seven PSI factors (significance denoted as $p < .05^*$, $p < .01^{**}$, $p < .001^{***}$; Osborne, Greene, & Immel, 2014)

The centering process is predominantly used to decrease excessive physiological arousal associated with performance anxiety (known as centering "down"), but can also be used to center "up" in some cases, to increase performance intensity and conviction. Interestingly, although statistically significant, the reduction in MPA (as measured by the Performance Energy subscale of the PSI) was moderate compared to changes on the other factors. This may be explained by the intervention's emphasis on the potential of high performance energy to facilitate better performance, rather than as a negative aspect of performance to be diminished in order to prevent a poor performance outcome.

In contrast to Kageyama's (Kageyama, 2007) inconclusive Julliard study using the centering process, this study demonstrated that conservatory musicians are able to understand their own optimal levels of performance energy (physiological adaptability) and apply strategies to redirect attention away from task-irrelevant to task-relevant musical cues (psychological adaptability) to improve their management of distressing performance anxiety and boost performance resilience (Osborne, Greene, & Immel, 2014).

Positive Doing

In attempting to understand why everyone wants to be a winner, and yet so few people achieve their dreams, performance psychologist Phil Jauncey (Jauncey, 2010, 2012) drew from Rotter's (1954) Locus of Control theory to articulate a simple message "If it is to be, it is up to me" based on the 'Internal Locus of Control' principle in which everything a person does is his/her responsibility. Traditional approaches to performance anxiety are based on the premise that unless the mind (emotions and thoughts) are in a very good place ('psychological automaticity'), a performer cannot have optimal performance. Since even those with the best of emotional control cannot always regulate their mental states, this approach often fails. The *Positive Doing* method assumes that one, when wanting to perform well at an important event (audition, prestigious venue, radio or TV), *will* have mental doubts, fears and trepidation. Rather than alleviating the emotion, the performer can *embrace* it, and thus overcome its effects.

Excuses are for Losers

There are only four reasons why we fail to perform:

1. We do not know what to do.
2. We do not know how to do it.

3. We do not have the ability to do it.
4. We choose not to do it.

Assuming the performer does not have the first three, then poor performance is a choice. Since it is a choice, we can always deal with any situation confronted. There are two factors that lead to the choice to perform badly:

1. Many believe that "*If only*" things beyond their control are right, then they can perform well.
2. When 'under the pump' our mind will spontaneously want to self-sabotage.

To be a winner, Jauncey (2010) proposes *Six Steps for Success*, starting with two fundamental questions:

1. What am I doing now?
2. Is it working?

If yes, continue the behavior.

3. If not, change!

This sequence of questions places the responsibility for outcomes internally (within the individual). If performers want to perform better under pressure, they must be willing to critique (not criticize) themselves. If they are doing something correctly, they should continue. If not, they need to change. For example, "When I found that my performance was below par, I needed to focus on the fundamentals of my activity to get back into it" (Jauncey, 2010). Not only would we expect performing individuals to learn from their mistakes so that if the conditions repeated themselves, the results would be better, they are also more likely to alter their performance during the course of an event, and hence, improve their results.

Yet, Jauncey observes that most people prefer to act with an externally-driven locus of control, where responsibility for performance outcomes is attributed to factors outside our control, e.g., "I would have liked to have done well, but it was just one of those days" (Jauncey, 2010). The difference between winners and losers is that winners interpret poor performance as a learning experience, whereas losers interpret such experiences as excuses – "It wasn't my fault." Such explanations diminish efforts to problem solve and improve performance. This ethos is similar to Carol Dweck's work on motivation (Dweck, 2012), showing that successful individuals are characterized by a psychological *growth* mindset, which is based on the belief that one's qualities are cultivated through one's efforts, where anyone can change and grow through application and experience (i.e., *winners*, and those who have maximized their potential) (Dweck, 2010). Conversely, individuals with *fixed* mindsets define their qualities using global, all-or-nothing statements in which outcomes are a direct measure of their innate competence and worth. In Jauncey's terms, those individuals are classified as *losers*, which includes individuals with immense talent and skills who have the "*If only...*" mindset (Jauncey, 2010). Performers who defer to "*If only...*" statements, such as "If only my teacher had..." or "If only I had that lucky break..." demonstrate that they have minimal control over outcomes, and do not have the mindset to succeed. Those who accept responsibility and act on the basis of "*Only if...*" statements have a much greater chance of success, when coupled with discipline, effort, critical appraisal and subsequent refined improvements.

In the *Positive Doing* method, a further three steps are required in order to change from a passive, fixed "*If only...*" mindset, to a pro-active, winning "*Only if* I do..., can I perform well" mindset:

4. Be specific
5. Make a plan
6. Act on it

Be Specific

Without specific, positive articulations of desired performance outcomes, we are unable to observe outcomes that indicate a lack of improvement. Performance goals that are defined in the negative (i.e., "Do not...") focus the mind on what is not desired and thus encourages that outcome. Performance goals and learning outcomes must indicate the desired behavior (what to do), rather than the undesired behavior (what not do to).

Make a Plan

Implementation intentions that delineate when, where, and how specific strategies will be enacted provide direction and purpose to achieve desired performance goals (Gollwitzer & Sheeran, 2006). Evidence for the effectiveness of "*If... Then*" plans to enhance the rate of goal attainment is strong across a number of domains (Achtziger, Gollwitzer, & Sheeran, 2008; Webb & Sheeran, 2007).

Act on It

Positive thinking is insufficient because it can't be observed. Behaviors and constructive action in line with one's performance goal is needed in order to achieve success. Without objective evidence, it is impossible to assess how close one is to achieving his/her performance goals.

These six steps help performers to identify and utilize their "Plan A flags," which are the normal actions that cue successful performance, to boost their ability to perform optimally 'under the pump.' For example, performers may use the centering technique as a *Positive Doing* flag ("Plan A" pre-performance routine) to facilitate a good performance, in addition to other supported strategies such as "power posing" (Cuddy, Wilmuth, & Carney, 2012) and deliberate performance practice and simulation (Williamon, Aufegger, & Eiholzer, 2014; see also Osborne, 2016, for additional strategies for performers of all ages).

In 2013 the author conducted the first empirical study of Jauncey's approach with musicians. The *Power of Positive Doing* program at the Australian National Academy of Music (ANAM) had been run and attended by staff and students since 2011. Anecdotal evidence suggested that it improved their ability to cope with performance anxiety. Nineteen participants (out of twenty-three in the first year cohort) were randomly allocated to complete the program either as part of the Academy's Health and Wellbeing week activities at the start of the academic year (Group 1, March), or waited to participate in the second delivery of the program in early May (Group 2: wait-list control). This method enabled the research team to understand whether any improvements were a result of the performance psychology program rather than the normal course of study. The program was comprised of six sessions delivered by two performance psychologists. The first two sessions provided psychoeducation regarding the physiological basis of anxiety and physiological arousal management strategies. Two subsequent sessions covered Jauncey's *Positive Doing* method including the *Six Steps for Success*, 'flags' and the 'myth of performance anxiety' (that is, the moment you start acting un-anxiously, you are no longer anxious).

The third and fourth sessions were devoted to explaining the rationale and results of Jauncey's 40-item personality profiling system. This system generates individual personality profiles along four dimensions to develop an individual's understanding of helpful and maladaptive behavior in situations of high performance stress and the most appropriate method of acting on plans.[1]

Students completed self-report assessments of MPA using Kenny's Music Performance Anxiety Inventory (K-MPAI; Kenny, 2011), which took place immediately before and after the first program and two months after program completion. The 40-item K-MPAI was modified for the pre- to post-intervention design to exclude seven items relating to past history of anxiety, parental empathy and generational transmission of anxiety (items 5, 9, 20, 23, 27, 29, and 33).

The research participants were mainly string players, followed by percussion and brass, then harp, wind and piano students. The mean age of the participants was 22.56 years (SD = 1.98 years), and 58% were female. Examining pre- and post-program scores collapsed for both groups of students showed that the program significantly reduced performance anxiety [Pre: $M = 79.01$, $SD = 24.68$; Post: $M = 70.43$; $SD = 23.43$; $t(18) = 2.84$, $p < .05$; Figure 5]. Further examination by group and time showed a significant pre- to post-program reduction for the wait-list control group who undertook the program as a small group of 10, $t(9)=2.55$, $p < .05$. The first intervention group reductions in MPA were also maintained over time, with no change in mean scores from post-test to two-month follow-up, $t(8)=1.10$, $p > .30$.

1 Readers interested in finding out more about this approach in sporting contexts are encouraged to consult Chapter 12 of Leigh Matthews' 2013 autobiography *Accept the Challenge*, one of the Australian Football League's most outstanding coaches and awarded player of the 20th century who describes Jauncey's personality profiling system, as well as stellar Australian cricketer Matthew Hayden's 2010 autobiography *Standing My Ground.*

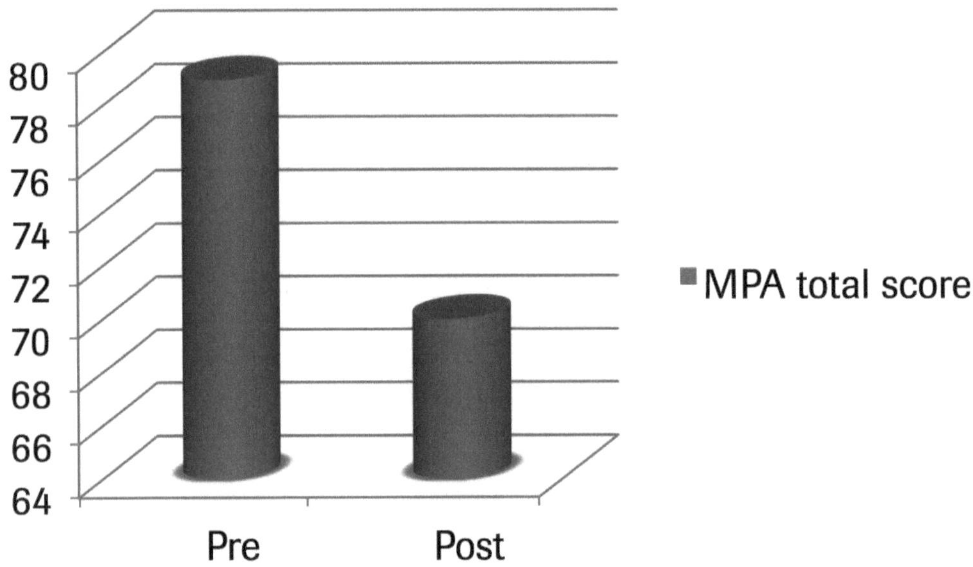

Figure 5: Music performance anxiety scale total score (modified K-MPAI) for all students immediately before and after participating in the program (significant at $p < .05$)

This study demonstrates that performance psychology strategies taught in a music performance professional training academy do reduce self-reported MPA, especially when delivered in a small group of 10. For those considering teaching performance psychology strategies in similar settings, it is important to remember that the method of delivery affects outcomes. Teaching in a small group format led to enhanced outcomes (significantly reduced MPA), which were not observed by students who participated in the larger group format.

Find What Works for You

The psychological skills presented in this chapter support the psychological well-being of performers, which is one of three factors which need to work in tandem in order to achieve performance excellence in stressful situations:

1. Competence (e.g., adequate skill development, practice, and task mastery);
2. Physical well-being (e.g., adequate nutrition, sleep, and fitness);

3. Psychological well-being (e.g., resilience and recovery from elevations in stress and anxiety through positive and realistic self-statements and realistic self-appraisal of performance quality) (Kenny & Ackermann, 2009; Williamon, 2004).

An individual's positive or negative interpretation of anxiety is mediated by their performance history, previous experience with elevated arousal levels, degree of skill or task mastery, self-confidence, self-belief in one's ability to execute the task correctly, and control over physical and cognitive anxiety symptoms (Gill, Murphy, & Rickard, 2006; Kenny, 2011; Rapee & Medoro, 1994). This means that performers need to identify, record, and examine their own triggers (or flags) to successful and unsuccessful performance outcomes. In so doing, they will create an individualized performance plan which includes flags for successful performance so that they can take constructive action that is appropriate to them, which takes into account the characteristics of various performance scenarios (e.g., amount of preparation, solo versus group performance, size and expertise of the audience, time of day, etc.). In all cases, performers can be helped by psychological strategies in pressure situations when they accept the following:

1. an optimal or peak performance (where performers think and feel they are performing at their best) is not a perfect performance. It is one where mistakes are fewest, and/or, minor problems are recovered from quickly (Williams, 2010). Musicians *do* make mistakes, and this does not necessarily mean performance failure;
2. poor performances provide valuable information on how to learn to improve performance in future events;
3. experiencing anxiety prior to, during, or after a performance is not inherently maladaptive and should not be avoided *per se*; it is in fact the natural response of our human nervous system providing us with alertness and energy to face any challenging situation; and,
4. anxiety may actually *enhance* performance.

This chapter has presented a program of research on mental skills training protocols which draw from the discipline of sports psychology and include techniques that integrate mind and body reactions under stress to enable a calm, focused, flexible and goal-oriented state of mind. It includes evidence-based strategies for performers to confidently *'control the uncontrollable'* spike in performance energy in high pressure situations to support them in executing an optimal performance. When performers are encouraged to understand, accept, and regulate the emotional and physiological symptoms that arise naturally when performing under pressure, they can be empowered to take constructive action and execute an optimal performance despite them.

References

Achtziger, A., Gollwitzer, P. M., & Sheeran, P. (2008). Implementation intentions and shielding goal striving from unwanted thoughts and feelings. *Personality and Social Psychology Bulletin, 34*(3), 381-393. doi: 10.1177/0146167207311201

American Psychological Association. (2011). *Defining the practice of sport and performance psychology*, Division 47 (Excercise and Sport Psychology) Practice Committee of the American Psychological Association. http://www.apadivisions.org/division-47/about/resources/defining.pdf

Appel, S. S. (1976). Modifying solo performance anxiety in adult pianists. *Journal of Music Therapy, 13*(1), 2-16.

Beckmann, J., Gröpel, P., & Ehrlenspiel, F. (2013). Preventing motor skill failure through hemisphere-specific priming: Cases from choking under pressure. *Journal of Experimental Psychology: General, 142*(3), 679-691.

Braden, A. M., Osborne, M. S., & Wilson, S. J. (2015). Psychological intervention reduces self-reported performance anxiety in high school music students. *Frontiers in Psychology, 6*(195), 1-9. doi: 10.3389/fpsyg.2015.00195

Brandon, C., & Ivens, C. (2009). *Thinking skills for peak performance: Student workbook and Coaches manual.* South Yarra: Macmillan Education Australia.

Britsch, L. (2005). Investigating performance-related problems of young musicians. *Medical Problems of Performing Artists, 20*(1), 40-47.

Brooks, A. W. (2014). Get excited: Reappraising pre-performance anxiety as excitement. *Journal of Experimental Psychology: General, 143*(3), 1144-1158. doi: 10.1037/a0035325

Cuddy, A. J. C., Wilmuth, C. A., & Carney, D. R. (2012). The benefit of power posing before a high-stakes social evaluation. *Harvard Business School Working Paper, 13-027.*

Deen, D. R. (2000). Awareness and breathing: Keys to the moderation of musical performance anxiety. *Dissertation Abstracts International, A (Humanities and Social Sciences), 60*(12-A), 4241.

Dweck, C. S. (2010). Even geniuses work hard. *Educational Leadership: Giving Students Meaningful Work, 68*(1), 16-20.

Dweck, C. S. (2012). *Mindset: How you can fulfil your potential.* London: Constable & Robinson.

Fehm, L., & Schmidt, K. (2006). Performance anxiety in gifted adolescent musicians. *Journal of Anxiety Disorders, 20*(1), 98-109.

Gill, A., Murphy, F., & Rickard, N. S. (2006). A preliminary examination of the roles of perceived control, cortisol and perceptions of anxiety in music performance. *Australian Journal of Music Education, 1*, 32-47.

Gollwitzer, P. M., & Sheeran, P. (2006). Implementation intentions and goal achievement: a meta-analysis of effects and processes. *Advances in Experimental Social Psychology, 38*, 69-119.

Greene, D. (2002). *Performance success: Performing your best under pressure.* New York: Routledge.

Greene, D. J. (2012, 17 January 2014). *11 strategies for audition and performance success: A workbook for musicians.* Retrieved 14 February, 2013, from http://psi.dongreene.com/

Gucciardi, D. F., & Dimmock, J. A. (2008). Choking under pressure in sensorimotor skills: Conscious processing or depleted attentional resources? *Psychology of Sport and Exercise, 9*(1), 45-59.

Hancock, P. A., & Ganey, H. C. N. (2003). From the inverted-U to the extended-U: The evolution of a law of psychology. *Journal of Human Performance in Extreme Environments, 7*(1), 5-14. doi: 10.7771/2327-2937.1023

Hancock, P. A., & Warm, J. S. (1989). A dynamic model of stress and sustained attention. *Human Factors, 31*(5), 519-537.

Hanton, S., Thomas, O., & Maynard, I. (2004). Competitive anxiety responses in the week leading up to competition: The role of intensity, direction and frequency dimensions. *Psychology of Sport and Exercise, 5*(2), 169-181.

Jackson, S. A., & Csikszentmihalyi, M. (1999). *Flow in sport.* Champaign, IL: Human Kinetics.

Jauncey, P. (2010). *Understanding ourselves and others.* Brisbane, Qld: CopyRight.

Jauncey, P. (2012). *The Power of Positive Doing.* Retrieved November 18, 2012, from http://www.philjauncey.com.au/

Jones, G., Swain, A., & Hardy, L. (1993). Intensity and direction dimensions of competitive state anxiety and relationships with performance. *Journal of Sports Sciences, 11*(6), 525-532.

Kageyama, N. J. (2007). *Attentional focus as a mediator in the anxiety-performance relationship: The enhancement of music performance quality under stress.* (Doctor of Philosophy thesis), Indiana University.

Kenny, D. T. (2011). *The psychology of music performance anxiety.* Oxford: Oxford University Press.

Kenny, D. T., & Ackermann, B. (2009). Optimizing physical and psychological health in performing musicians. In S. Hallam, I. Cross, & M. Thaut (Eds.), *Oxford Handbook of Music Psychology* (pp. 390-400). Oxford, UK: Oxford University Press.

Kenny, D. T., Driscoll, T., & Ackermann, B. (2014). Psychological well-being in professional orchestral musicians in Australia: A descriptive population study *Psychology of Music, 42*(2), 210-232. doi: 10.1177/0305735612463950

Kim, J. K., & Diamond, D. M. (2002). The stressed hippocampus, synaptic plasticity and lost memories. *Nature Reviews Neuroscience, 3*(6), 453-462.

Krane, V., & Williams, J. M. (2010). Psychological characteristics of peak performance. In J. M. Williams (Ed.), *Applied sport psychology: Personal growth to peak performance* (6th ed., pp. 169-188). New York: McGraw-Hill.

Mansberger, N. B. (1988). *The effects of performance anxiety management training on musician's self-efficacy, state anxiety and musical performance quality.* (Master of Music thesis), Western Michigan University, Kalamazoo, Michigan.

Martin, A. J. (2008). *Motivation and Engagement Scale Music (MES-M) Test User Manual.* Summer Hill: Lifelong Achievement Group.

Mesagno, C., & Mullane-Grant, T. (2010). A Comparison of Different Pre-Performance Routines as Possible Choking Interventions. *Journal of Applied Sport Psychology, 22*(3), 343-360.

Milton, J., Solodkin, A., Hlustik, P., & Small, S. L. (2007). The mind of expert motor performance is cool and focused. *NeuroImage, 35,* 804-813.

Mornell, A. (2002). *Lampenfieber und Angst bei ausübenden Musikern* (Vol. 14). Frankfurt am Main: Peter Lang.

Mornell, A. (2016). Shining a spotlight on stage fright: Removing the shadows of fear. In W. Mastnak (Ed.), *Modern piano teaching. A comprehensive and psychologically based guideline* (pp. 210-222). Shanghai: Shanghai Conservatory of Music Press

Münte, T. F., Altenmüller, E., & Jäncke, L. (2002). The musician's brain as a model of neuroplasticity. *Nature Reviews Neuroscience, 3*(June), 473-478.

Nideffer, R. M., & Sagal, M.-S. (2001). Concentration and attention control training. In J. M. Williams (Ed.), *Applied sport psychology: personal growth to peak performance* (4th ed., pp. 312-332). Mountain View, CA: Mayfield.

Osborne, M. S. (2008). *Music performance anxiety in young musicians: Conceptualisation, phenomenology, assessment and treatment.* (Doctor of Philosophy thesis), The University of Sydney, Australia.

Osborne, M. S. (2013). Maximising performance potential: The efficacy of a performance psychology program to reduce music performance anxiety and build resilience in adolescents. In A. Williamon & W. Goebl (Eds.), *Proceedings of the International Symposium on Performance Science 2013* (pp. 303-310). Brussels, Belgium: Association Européenne des Conservatoires.

Osborne, M. S. (2016). Building performance confidence. In G. E. McPherson (Ed.), *The child as musician: A handbook of musical development* (2nd ed., pp. 422-440). Oxford: Oxford University Press.

Osborne, M. S., Greene, D., & Immel, D. (2014). Managing performance anxiety and improving mental skills in conservatoire students through performance psychology training: a pilot study. *Psychology of Well-Being, 4*(18), 1-17. doi: 10.1186/s13612-014-0018-3

Osborne, M. S., & Kenny, D. T. (2005). Development and validation of a music performance anxiety inventory for gifted adolescent musicians. *Journal of Anxiety Disorders, 19*(7), 725-751. doi: 10.1016/j.janxdis.2004.09.002

Osborne, M. S., Kenny, D. T., & Cooksey, J. (2007). Impact of a cognitive-behavioural treatment program on music performance anxiety in secondary school music students: A pilot study. *Musicae Scientiae, 11*(2 Special Issue), 53-84. doi: 10.1177/10298649070110S204

Rapee, R. M., & Medoro, L. (1994). Fear of physical sensations and trait anxiety as mediators of the response to hyperventilation in nonclinical subjects. *Journal of Abnormal Psychology, 103*(4), 693-699.

Reitman, A. D. (2001). The effects of music-assisted coping systematic desensitization on music performance anxiety. *Medical Problems of Performing Artists, 16*(3), 115-125.

Roland, D. J. (1994). The development and evaluation of a modified cognitive-behavioural treatment for musical performance anxiety. *Dissertation Abstracts International, 55*(5-B), 2016.

Sweeney, G. A., & Horan, J. J. (1982). Separate and combined effects of cue-controlled relaxation and cognitive restructuring in the treatment of music performance anxiety. *Journal of Counseling Psychology, 29*(5), 486-497.

Wardle, A. (1975). Behavior modification by reciprocal inhibition of instrumental music performance anxiety. In C. K. Madsen, C. H. Madsen, & R. D. Greer (Eds.), *Research in music behavior: modifying music behavior in the classroom* (pp. 191-205). New York: Teachers College Press.

Webb, T. L., & Sheeran, P. (2007). How do implementation intentions promote goal attainment? A test of component processes. *Journal of Experimental Social Psychology, 43*(2), 295-302. doi: 10.1016/j.jesp.2006.02.001
Williamon, A. (Ed.). (2004). *Musical excellence: Strategies and techniques to enhance performance*. Oxford: Oxford University Press.
Williamon, A., Aufegger, L., & Eiholzer, H. (2014). Simulating and stimulating performance: introducing distributed simulation to enhance musical learning and performance. *Frontiers in Psychology, 5*(25), 1-9. doi: 10.3389/fpsyg.2014.00025
Williams, J. M. (Ed.). (2010). *Applied sport psychology: Personal growth to peak performance* (6th ed.). New York, NY: McGraw-Hill.
Yoshie, M., Kudo, K., Murakoshi, T., & Ohtsuki, T. (2009). Music performance anxiety in skilled pianists: Effects of social-evaluative performance situation on subjective, autonomic, and electromyographic reactions. *Experimental Brain Research, 199*(2), 117-126.

Relating Pianists' Efficiency in Memorization to Strategy Use and Experience

Memory is, probably next to language, the most researched topic in psychology. However, due to the much smaller number of music researchers compared to scholars in general psychology and education, studies on musical memory are less frequent on the whole (see Chaffin, Demos, & Logan, in press, for a review). Musical memory is interesting for various reasons: we remember songs, we identify redundancy in music, and a lot of solo music is performed from memory in the classical domain. This performance tradition emerged in the 18th and 19th century, possibly as a by-product of intense preparation of concerts by performers such as Franz Liszt, Hans von Bülow, and Clara Schumann (cf. Bitzan, 2010). Such preparation became necessary for those who were not exclusively performing their own works but were playing what came to be known as repertoire – a canonized set of pieces.

Two hundred years later the now-established performance tradition has led to the audience's expectation of hearing a well-known piece played note-perfectly by a concert musician. It is this expectation that scares performers and those who aspire to become concertizing musicians. From our experience – and probably many other colleagues' experience and anecdotal evidence – memory problems are the most frequent challenge during performance, and the fear of encountering such problems may be the most important reason for students' stage fright. While audiences (and jurors) only witness the result of months-long preparation for memorized performance, it is unclear how students prepare for it, which strategies they are using, and if their success is related to the type of strategies they use. And it is also unclear whether the ability to memorize music might be subject to substantial individual differences among musicians, which would explain certain problems and experiences during performance preparation. As Chaffin et al. noted, knowledge on memorization is

> conveyed from teacher to student through an apprenticeship system that makes it unavailable for systematic analysis. Memorization is viewed as an individual and mysterious process [...]. This is regrettable and unnecessary [...] (in press, p. 568).

Hence, research on memorization strategies is of practical importance (also Hallam, 1997, p. 87).

This paper will address three issues related to the topic. The literature reviewed centers around strategies for memorizing, individual differences, and a model of expert musical memory. In the study, we[1] were interested in first establishing which memorizing strategies would be used by experts to prepare for performance; second, how extensive the individual difference might be in memorizing ability; and lastly, whether superior memorizing ability was related to the use of specific strategies or memorizing experience in the course of music training.

Performance and Memorization

Most young musicians experience memorization as an apparently uncomplicated process; the piece simply becomes memorized without conscious effort. Although incidental memorization that happens during learning is experienced by the performer as unconscious, it proves unreliable under performance pressure (Chaffin, Imreh, & Crawford, 2002: see chapter 3). And only "[d]eliberate memorization transforms the motor and auditory chains created while learning the piece by making them content addressable" (Chaffin et al., in press, p. 559). Granted, "[m]emory failures are inevitable [...]. The important thing is to recover gracefully [...]. Most of the time, the audience is not even aware of the mistake" (Chaffin et al., in press, p. 560).

Musicians use a number of strategies to develop elaborate mental mechanisms when deliberately preparing music for memorized performance. The most developed theory, founded on the concept of skilled memory theory (Chase & Ericsson, 1982; Ericsson & Kintsch, 1995), stems from Chaffin and collaborators who have collected evidence from music performers over the past decade.

When memorizing music, performers can rely on schemas which consist of basic knowledge arrangements representing generic memories (e.g., what an Alberti bass accompaniment looks like). These are then filled out constructively using multiple constraints (Chaffin et al., in press, p. 561). Therefore, the more pianists know about a genre, style, and compositional rules, the more they will be able to construct an adequate representation of the memorized stimulus. Furthermore, previously played pieces will provide additional information and procedural knowledge to facilitate

1 Elisabeth Ott is credited for helping in resurrecting the data, which was originally collected in the 1990s, for use in a presentation at the 2012 *Art In Motion* symposium in Munich.

learning of novel material. The construction relies on different types of memory: auditory, motor, structural, emotional, visual, and linguistic (p. 561; cf. also Gerling & Santos, 2015).

The least productive mechanism, however, is associative chaining (sometimes referred to as "forward chaining"), whereby one motor or auditory chunk is used to cue the subsequent motor chunk which in turn cues the next. This resembles a little chain of train wagons, each one connected to the preceding by a single place of contact. If one connector fails, all following wagons (chunks) become inaccessible. Informal observation and questioning of music students reveals numerous episodes in which a piece was played from memory, and upon a memory failure, had to be repeated from the beginning in order to be continued beyond the problematic spot. Often, the score had to be consulted to continue, or in the absence of a score, the young musician left the stage, deeply ashamed and often traumatized for the future[2].

Temporary changes in ability to perform from memory arise from situational effects, such as stage fright (Hallam, 1997, p. 93; Mornell, 2002); memory failures and even complete blackouts were reported. Especially during puberty, some high-achieving young (often female, perfectionist) musicians experience high music-performance anxiety and ensuing memory problems (Osborne, 2015), which may be related to an altered way they conceptualize (represent) music, as well as an increased self-awareness when performing in front of an audience. However, one of Hallam's subjects reported improvement with repeated occasions for memorized performance, which is in line with most conservatoire teachers' experience. Further constraints on memorization are imposed by the amount of information (i.e., number of notes) to be memorized or its complexity (i.e., compositional structure) (e.g., Lehmann & Ericsson, 1998; Mishra, 2006; Rubin-Rabson, 1940).

Hallam (1997, p. 88) undertook a phenomenographic study on how experts and novices experienced learning and performing from memory. 22 professional and 55 novice musicians were interviewed with a semi-structured protocol. The interviews were transcribed and consensually coded by three raters. An "automated process" and a separate "cognitive process" entailing musical analysis of some sort were identified for the experts along with combined approaches. Their strategies were con-

2 It happened once to one of the authors of this paper (ACL) as a child.

text-dependent: for example, longer and shorter pieces required different approaches. On the whole, memorization seemed to be an individual and varied process, but the unconscious and cognitive (deliberate) components mentioned earlier were invariably found among the professional musicians.

The current standard model of expert performers' musical memory is the one first described by Chaffin, Imreh, and Crawford (2002). They explored a professional pianist's (Gebriela Imreh) memorizing process, describing the mental representations she arrived at when learning a piece of music. At the core of the model are the so-called "performance cues" that act as retrieval cues during memorized performance (also Chaffin, Lemieux & Chen, 2007; Lehmann & Chaffin, 2008; Chaffin et al., in press; Williamon & Valentine, 2002). They emerge in the learning process and are three-fold: basic cues (technical issues), interpretive cues (musical gestures and dynamics), and expressive cues (musical expression and character descriptions). When performing from memory, the musician sequentially uses these performance cues which act as milestones or signposts and facilitate retrieval. For these cues, features of the content are used to access the content, and part of the content can then be used to access other parts. This type of distributed memory is "able to work around errors by reconstructing information that may have been hidden or scattered throughout the system" (Gerling & Santos, 2015, p. 3). Whereas basic performance cues dominate in the beginning of learning a piece of music, they give way to cues related to structural aspects of the music (e.g., Williamon & Valentine, 2002). Performance cues exist in a hierarchical structure and can be accessed individually if need be.

Strategies for Memorizing

All famous music pedagogues have their shared knowledge about memorizing, for example, the pianists Leimer and Gieseking (1932/1972, cf. Lehmann, 1997; cf. Bitzan, 2010, p. 109ff for further examples). They describe – in musicians' terms – a multimodal coding strategy of a piece of music, mainly in terms of its compositional structure. They call this method "visualization." Interestingly, they advocate its use away from the instrument prior to actually playing the piece of music, presumably to prevent meaningless incidental memorization. Often music educators borrow analogies from the visual domain to talk about memorization methods, and their insights are conveyed in a recipe-type format (cf. Lehmann, 1997, p. 147).

Many studies have reported on the use of memorization strategies in musicians (see Mishra, 2007, for a review). For instance, in Bitzan's (2010) small online survey, participants reported the "repeating of short sections with score" (49%), "alternating between playing from score, then from memory" (45%), "short glances at score between sections played from memory" (56%), and "deliberate memorizing of short segments" (71%). In her naturalistic observation, Mishra (2002, cit. in 2011, p. 61) found four types of memorizing strategies at the macroscopic level: holistic (playing through the entire piece), segmented (working on segments of the piece), serial (starting at the beginning until an error occurred, then returning to the start), and additive (memorizing a segment and increasing its length successively). It appeared that participants using the holistic or additive strategies were more efficient memorizers than those using the segmented or serial strategy. Not all musicians are aware of the existence of memorizing strategies, and some report using none (38% in Bitzan, 2010; 22% in Mishra, 2007, cit. in 2011, p. 66). In a subsequent experiment with random assignment to the four conditions mentioned above, Mishra (2011) studied the efficiency and stability of retention of a 16-bar piece. The results showed the holistic or additive strategies to be more effective. Furthermore, subjects who were habitually using the strategy they had been assigned performed no better than the others.

As mentioned previously, musicians code music in different representations which can be visual, aural, kinesthetic, etc. (Chaffin et al., in press; Hallam, 1997; Mishra, 2007), and there may be individual differences in the use of such representations that contribute to memorizing efficiency. Learning style preferences (visual, aural, kinesthetic), also referred to as perceptual learning modalities measured with the Learning Style Test (LST) and the Vark Questionnaire, were weakly related to a researcher-developed Musical Memorizing Inventory (MMI, Mishra, 2007). Respondents tended to report mixed approaches, with visual LST and MMI preferences showing the highest association ($r[77] = .34$). Similarly, Hallam (1997, p. 95) states that "[t]hose musicians reporting the greatest confidence in performing from memory appeared to adopt multiple coding, both automated and analytic. This approach would minimize retrieval failure, enabling the utilization of a number of different retrieval cues." However, in their sample of undergraduate piano students, Gerling and

Santos (2015) found the use of kinesthetic and auditory cues to be moderately negatively correlated. Although different strategies exist and some may be more effective than others, there may be confounding variables that influence the use of strategy, such as experience with memorization, personal preferences, and structural aspects of the piece to be memorized (e.g., tonal vs. atonal).

Individual Differences

It is unclear how large the individual differences in music memorization ability may be among advanced players. Whereas short term-memory is fast, long-term memory where the music performances are stored takes longer to access, is content-addressable, and requires specific retrieval cues. Overcoming this bottleneck is the main challenge that musicians face when memorizing and performing from memory. Spectacular feats such as those of autistic savant Derek Paravincini, who can memorize a rather complex piece after a few or even a single presentation (Ockelford, 2007), mark the outer bounds of what human memory is capable of processing. Not only can Derek memorize almost instantaneously, but he is also able to transpose the piece to a different key which requires extensive remapping of pitch-to-key-to-finger.[3] One can view Derek's memorizing ability as an extreme case of experts' skilled memory, as it is restricted to a domain and its typical demands, is hierarchically organized, and emerges as a result of intense training and practice (see Chase & Ericsson, 1982; Ericsson & Kintsch, 1995). At the other extreme, there are persons who encounter great difficulties memorizing new material such as patients with Alzheimer dementia or with traumatic brain injuries.

Most musicians probably fall into the broad category between those extremes with some having more and less ease at memorizing. They "often decide whether they are 'good' or 'poor' memorizers based on a handful of salient experiences or informal conversations rather than a systematic observation and comparison. There is no doubt that some musicians memorize faster than others" (Mishra, 2006, p. 1184) – which she calls "efficiency," a term we will adopt. And in a footnote Mishra continued that she "found that university instrumentalists required between 9 and 100 minutes to memorize the same 36-bar exercise" (p. 1184). Large variability was also reported in Mishra (2011). Hallam's (1997, p. 92) par-

3 Some YouTube videos bear witness to his impressive feats and can easily be found by searching for his name.

ticipants had explicit conceptions about their abilities, such as "I've always had a memory . . . reliable . . . I still do," "I was never very good at memorizing," or "I'm hopeless, absolutely hopeless." Here, as elsewhere, no performance data were collected to corroborate the subjective claims.

Lehmann and Ericsson (1997; cf. also Lehmann, 1997) reported a laboratory experiment in which advanced pianists memorized two short pieces of music, each in a separate session. The authors used a standardized procedure (same as described below in the method section) with as many trials as necessary to memorize the music. The number of trials to criterion between the two pieces was highly correlated (.76), suggesting a generalized ability to memorize. However, individual differences were large, with one person requiring less than five trials per piece and another needing over 20. Subsequently, subjects were asked to perform tasks using the memorized material, such as playing considerably faster/slower, performing hands separately, and even transposing the piece to a different key. Contrary to what one might have suspected, faster (more efficient) memorizers were more accurate on the transformed performance tasks ($r >= .70$). Since the memorization strategy was imposed on the experiment and habitual strategy use was not assessed, it was unclear if efficiency was related habitual strategy use.

In the current study, we replicated findings from Lehmann and Ericsson (1997) regarding individual differences in efficiency, gaining insights into strategy use of expert pianists, and assessed the amount of memorized repertoire during lifetime as an indicator of memorizing experience. Since this paper focuses on surveys of memorizing strategies, the memorization experiment will only be briefly touched upon. In the present study the dependent variable was memorizing ability (efficiency).

Memorization Study

Subjects

Twenty advanced pianists between the ages of 19 and 44 took part in the study (11 female; age $M = 26.8$, $SD = 8.5$; *median* 24). They were enrolled in piano performance degree programs at various levels. Participants were recruited by word of mouth and posted flyers; they signed an informed consent form prior to the experiment/interview, and they were remunerated for their cooperation.

Biographical Interviews and Memorizing Strategies

In the course of an hour-long structured biographical interview, participants were asked about their experiences with memorizing in general and their music memorizing strategies in particular. The latter was also assessed quantitatively using a survey that contained 24 strategies compiled from the literature and from exploratory pilot interviews with different musicians (see Figure 1, p. 127). Participants responded on a 10-point scale as to whether they used the given method to prepare for a public performance (1 = "I never use this method"; 10 = "I always use this method"). Durations of deliberate practice for each year were calculated from weekly practice estimates (cf. Ericsson, Krampe, & Tesch-Römer, 1993, for details on the method), and other training variables were computed.

Assessing Memorization Efficiency (Ability)

Two short pieces, composed especially for the experiment, one in *e*-minor and one in *G*-Major, were memorized by the participants, each memorized in a separate experimental session. The pieces were eight measures long and consisted of roughly 70 notes, divided equally between the left and right hand. The style of music was tonal and consisted of a typical four-part, slightly polyphonic arrangement that could serve as an accompaniment to a solo voice. The experimental procedure, entitled "accompanying paradigm," was similar to that developed by the authors in the context of sight-reading research (Lehmann & Ericsson, 1993). It allowed us to pace the performance in an ecologically valid manner akin to an accompanying situation with an instrumental music soloist. During the memorizing session, a score of the music was presented on a computer display in front of the musician who was seated at a high quality MIDI keyboard (88 weighted keys). After a short count-in (MIDI metronome clicks), the pianists performed the score along with a pacing solo voice which was presented over loud speakers. Next, the screen went blank, the count-in started over, and the musician performed the music from memory, again paced by the solo voice. This alternating procedure was repeated until the criterion of a note-perfect performance from memory was achieved. The number of attempts was viewed as a measure of memorizing ability (efficiency) with smaller numbers representing superior memorizing efficiency. For ease of interpretation, we used $-m_{trial}$ such that lower numbers indicated higher efficiency. The descriptive statistics on the trials for the *e*-minor piece showed a similar number of attempts to memorize of $M = 31.0$ ($SD = 11.64$), $SE = 2.6$ (*Range* 15 - 49); for the *G*-major piece subjects required a mean number of attempts of $M = 33.0$ ($SD = 14.6$), $SE = 3.2$ (*Range* 9 - 55). The correlation of memorization attempts between pieces was $r(18) = .73$, $p < .01$. Thus, the pieces were of similar difficulty and the measure was reliable.

Results

The first research question concerned the existence of individual differences in memorizing efficiency. For this, we summed the number of trials for the two pieces ($M = 64.05$, $SE = 5.46$, *95% CI* [52.61, 75.48]). When performing a median split on the efficiency variable and using a t-test to assess group differences ($T[18] = 7.85$, $p < .01$), the large individual differences in efficiency could be expressed in terms of an effect size: 3.5 (*95%CI* [3.01, 4.04]).

The use of strategies may be diverse, and good and weak memorizers might be using different strategies in their daily routines. To analyze their use of strategies, we computed descriptive statistics (see Figure 1). The most used strategies were "Thinking through the music in general terms away from the instrument" (Strat 2), "Practice specific entry points" (Strat 18), and "Playing slowly from memory" (Strat 4), while the least used ones were "Improvise on the piece or on sections of the piece" (Strat 11), "Play from memory faster than the expected performance tempo" (Strat 3), and "Listen to recordings and try to play back by ear after hearing" (Strat 21). Three of the variables (2, 11, 18) were transformed to reduce skewness. Next, we correlated strategy with efficiency of memorization. The resulting correlations were small to medium size (-.35 to .38, n.s.). The only exception was Strat 2 ("Think through the music in general terms in your head away from the instrument"), $r_s(18) = .60$, $p < .05$. This same strategy was also moderately correlated with "Listen to recordings and try to play back by ear after hearing" (.56, $p < .025$), "Anticipate how the music feels under the hand while playing (slowly) from memory" (.57, $p < .01$), and "Work with the score away from the instrument" (.54, $p < .025$). Those who used Strat 2 tended to be more efficient, and despite the fact that a factor analysis of the 24 strategies was impossible due to the small N, the patterns of correlations for Strat 2 suggests a mental occupation with the music, along with an awareness of their mental representations.

The interview had focused on training indicators, such as memorizing experience and piano practice durations. The next analysis investigated the relation between memorizing efficiency and training parameters. An estimated percent of memorized repertoire was tabulated for every year since the beginning of piano instruction, as were piano practice durations. As there was a considerable age range among participants, but subjects had all started to play as children (age $M = 6.1$, $SD = 1.97$), practice times and memorized repertoire were calculated for the times up to the 15th and 20th birthday, respectively. Lifetime piano practice ($r[18] = .27$), piano practice up to age 15 (.13), or practice up to age 20 (.10) was not reliably related to memorizing efficiency, whereas correlations were slightly higher for current memorized repertoire (.31) and percent memorized repertoire up to age 15 (.39) and 20 (.37).

Performing a stepwise regression predicting efficiency from memorizing experience at age 15 and age 20 resulted in $F(1,18) = 3.2$, $p = 045$ (one-tailed) with an *R2* of 15% and $F(1,18) = 2.9$, $p = 053$ (one-tailed) with an *R2* of 14%, respectively. By performing a median split on memory efficiency, we could estimate the magnitude of group differences. For age 15, the difference was 25.74 percent points and corresponded to an effect size of $d = .78$ (95%CI[.56, .99]) with more efficient memorizers having memorized more percent of repertoire ($M = 52.6$, $SD = 30.47$) than less efficient ones ($M = 32.3$, $SD = 21.0$). For age 20, the difference was 21.76 points with an effect size of $d = .76$ (*95%CI*[.54, .97]); again, more efficient memorizers had more ($M = 61.3$, $SD = 24.96$) repertoire memorized than less efficient ones ($M = 44.8$, $SD = 18.0$).The results were consistent with our hypothesis, even though the 20 year variable just missed the significance level due to our small sample size. Potentially large effect sizes were present.

Discussion and Conclusions

We found unexpectedly large individual differences among advanced musicians - a finding which supports Mishra's (2006) observations. While novice-expert comparisons naturally find large effects, such individual differences among experts come as quite a surprise. In everyday life they often go unnoticed, because the audience/teacher only sees the end result - a memorized performance. How much time and effort were invested remains a well-kept secret; according to our preliminary analysis, the effect size between strong and weak memorizers is at least between three and four standard deviations. Transferred to the logic of IQ measurements, we are looking at a score of 70 versus 130! Only an experiment such as the one discussed above can uncover such differences and quantify them.

Superior memorization does not seem to be connected to the use of a single strategy. In this, our results are consistent with those of other authors. Unfortunately, our survey did not distinguish between acquisition of the initial incidental memory trace and the secondary deliberate memorizing process. Yet, our experimental results suggest that consistent, long-term use of strategies which require working with the score on a conceptual level seemed to be more beneficial for memorizing than largely procedural strategies (which may still allow the formation of incidental motor memory traces).

We found that estimates of previous memorizing experience were positively related to memorizing efficiency, such that participants who had memorized more music in their childhood and adolescence did better in the experiment. Our results do not explain much variance (around 15%), which may be due to the rather vague indicator (the percent of memorized repertoire for a given year), and our subjects' difficulties in assessing the facts retrospectively. Unlike daily practice which tends to be regular, music memorization varies depending on the stage of learning a piece. While quantity of piano practice was unrelated to efficiency, specific memorization experience was; this is consistent with similar results regarding accompanying experience and sight-reading skills (Lehmann & Ericsson, 1993) and more general results on the effect of task-specific accumulated practice on musical achievement (Platz, Kopiez, Lehmann & Wolf, 2014; aggregated effect size of $r_c = .61$). However, mere experience is not "deliberate practice" as defined by expertise theory (cf. Ericsson, Krampe & Tesch-Römer, 1993; Platz et al., 2014, for a recent discussion); better indicators of deliberate memorizing training have to be found in future research. Also, our results are of correlational nature, and we cannot rule out that participants who memorized more efficiently as children engaged more in this activity. However, our interview partners often attributed their earlier efforts to demands the teacher imposed on them ("My teacher made me memorize"), not to their own proclivity.

How do we combine incidental memory, deliberate memorization, memorizing strategies, and the existence of performance cues? We suggest that incidental memorization can include basic performance cues but that these are largely embedded in an associative chaining procedure which is insufficient for stage performance. The conscious memorization process can take advantage of those performance cues that emerge while learning the piece. However, once the performer deliberately memorizes using more conceptual strategies, performance cues (including the basic ones) become content-addressable and part of the imaginary landscape (mental map) of the piece. Ultimately, there will be times during a performance when the performer switches on "the auto pilot" and directs the motor activity from the highest hierarchical level of control (e.g., "now comes the second movement"). In case of a memory failure, the performer can move down the command line to the closest performance cue or jump to the next "jump point," most likely also a performance cue (cf. Chaffin et al., 2002). When performing under pressure, an expert memorizer then uses the performance cue that is strategically most useful – or accessible (Ginsborg, Chaffin & Demos, 2012).

What can we learn from the exploration of memorization methods and efficiency for music instruction? Teachers are probably unaware of the large individual differences of their students' memorization efficiency. This lack of awareness must result in tremendous pressure felt by some students, for instance, when the teacher expects a piece memorized by the next lesson. Students who do not memorize easily might "fake" memorized performance by relying more on the incidental rote memory developed during learning (instead of deliberate memorization), which creates problems in stressful situations. Finding that the use of conceptually oriented strategies proved beneficial in our study should encourage students to analyze their pieces, work with the score and create a vivid – mostly non-motor – mental map of the piece. The intervention study by Gerling and Santos (2015) suggests promising results. Since early memorizing experience was associated with efficiency of memorization, teachers should foster memorizing skills beyond incidental memory in beginners. Learning about strategies might also prevent catastrophic memory failures in childhood/adolescence, which in turn would result in an improvement in the beliefs that music students hold regarding their own ability to memorize.

In conclusion, in addition to simply learning the piece, performing memorized music under pressure requires specific preparation which the learner has to engage in deliberately. The general psychology literature has amply documented the incredible trainability of memory. There is absolutely no reason for young musicians to be afraid of performing from memory if teachers prepare them well, which will happen, if teachers question their own learning biography and assumptions.

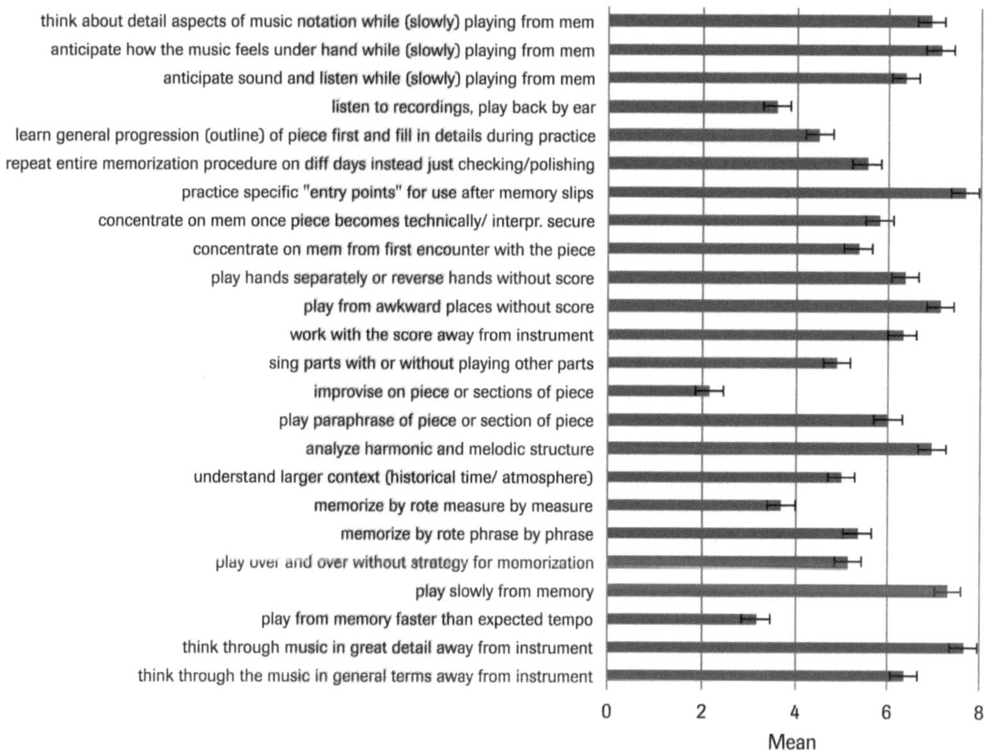

Figure 1. Mean ratings for memorizing strategies (higher numbers signal more frequent use). Error bars indicate 95% confidence intervals ($N = 20$).

References

Bitzan, W. (2010). *Auswendig lernen und spielen.* Über das Memorieren in der Musik. Frankfurt: Lang.
Chaffin, R., Lemieux, A. F., & Chen, C. (2007). 'It is Different Each Time I Play.' Variability in highly prepared musical performance. *Music Perception, 24*, 455–472.
Chaffin, R., Imreh, G., & Crawford, M (2002). *Practicing perfection: Memory and piano performance.* Mahwah NJ: LEA.
Chaffin, R., Demos, A. P., & Logan, T. (in press). Performing from memory. In S. Hallam (Ed.), *Oxford Handbook of Music Psychology* (chapter 35). Oxford: Oxford University Press.
Chase, W. G., & Ericsson, K. A. (1982). Skill and working memory. In G. H. Bower (Ed.), *The psychology of learning and motivation*, vol. 16 (pp. 1–58). New York: Academic Press.
Ericsson, K. A., & Kintsch, W. (1995). Long-term working memory. *Psychological Review, 102,* 211-245.
Ericsson, K. A., Krampe, R., & Tesch-Römer, C. (1993). The role of deliberate practice in the acquisition of expert performance. *Psychological Review, 100*, 363-405.
Gerling, C. C., & Santos, R. A. (2015). How do undergraduate piano students memorize their repertoire*? International Journal of Music Education.* DOI: 10.1177/0255761415619427
Gieseking, W., & Leimer, K. (1932/1972). *Piano technique.* New York: Dover.
Ginsborg, J., Chaffin, R., & Demos, A. F. (2012). Different roles for prepared and spontaneous thoughts: A practice-based study of musical performance from memory. *Journal of Interdisciplinary Music Studies, 6*(2), 201-231.Online accessed (1-6-2016) at http://www.musicstudies.org/JIMSvol6issue2PRINT/Ginsborg_JIMS_12060205.pdf
Hallam, S. (1997). The development of memorization strategies in musicians: Implications for education. *British Journal of Music Education, 14*, 87-97.
Lehmann, A. C., & Chaffin, R. (2008). Erinnern und Wiedererkennen: Auswendig- und Vomblattspiel [Remembering and recognizing: Playing from sight and memory]. In H. Bruhn, R. Kopiez & A. C. Lehmann (Eds.), *Musikpsychologie: Das neue Handbuch* (pp. 354-373). Reinbek: Rowohlt.
Lehmann, A. C. (1997). Acquired mental representations in music performance: Anecdotal and preliminary empirical evidence. In H. Jørgensen & A. C. Lehmann (Eds.), *Does practice make perfect* (pp. 141-164). Oslo: Norges Musikhogskole.
Lehmann, A. C., & Ericsson, K. A. (1993). Sight-reading ability of expert pianists in the context of piano accompanying. *Psychomusicology, 12*(2), 182-195.
Lehmann, A. C., & Ericsson, K. A. (1997). Expert pianists' mental representations: Evidence from successful adaptation to unexpected performance demands. In A. Gabrielsson (Ed.), *Proceedings of the 3rd Triennial ESCOM Conference* (pp. 165-169). Uppsala, Sweden: Uppsala University.
Lehmann, A. C. & Ericsson, K. A. (1998). Preparation of a public piano performance: The relation between practice and performance. *Musicae Scientiae, 2,* 69-94.
Mishra, J. (2006). Predicting memorization efficiency through compositional characteristics. In M. Baroni, A. R. Addressi, R. Caterina & M. Costa (Eds.), Proceedings of the 9th ICMPC, Bologna, Italy, August 22-26, 2006.
Mishra, J. (2007). Correlating musical memorization styles and perceptual learning modalities. *Visions of Research in Music Education, 9/10*, 1-9. Online accessed (12-31-2015) at http://www-usr.rider.edu/~vrme/v9n1/vision/Mishra%20Final.pdf
Mishra, J. (2011). Influence of strategy on memorization efficiency. *Music Performance Research, 4*, 60-71.

Mornell, A. (2002). *Lampenfieber und Angst bei ausübenden Musikern* [stage fright and anxiety in performing musicians]. Frankfurt: Peter Lang.
Ockelford, A. (2007). A music module in working memory? Evidence from the performance of a prodigious musical savant. *Musicae Scientiae* (Special Issue), 5–36.
Osborne, M. S. (2015). Building performance confidence. In G. E. McPherson (Ed.), *The child as musician: A handbook of musical development* (2nd ed.). Oxford: Oxford University Press.
Platz, F., Kopiez, R., Lehmann, A. C., & Wolf, A. (2014). The influence of deliberate practice on musical achievement: A meta-analysis. *Frontiers in Psychology - Cognition, 5*:646. doi: 10.3389/fpsyg.2014.00646
Rubin-Rabson, G. (1940). Studies in the psychology of memorizing piano music: III. A comparison of the whole and the part approach. *Journal of Educational Psychology, 31*, 460-476.
Williamon, A., & Valentine, E. (2002). The role of retrieval structures in memorizing music. *Cognitive Psychology, 44*, 1-32.

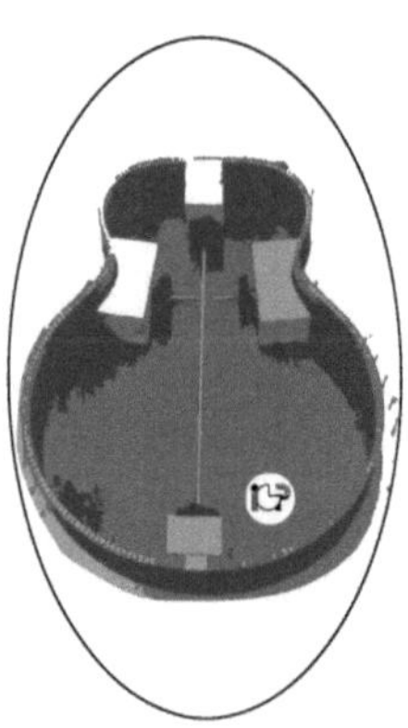

G. McPherson, P. Evans, E. Kupers, J. Renwick

Applying Self-Determination and Self-Regulation Theories for Optimizing Music Performance

The most prominent explanation for how musicians attain expert performance shows that they undergo 10,000 hours of highly structured and effortful *deliberate practice* across at least 10 years (Ericsson, Krampe & Tesch-Romer, 1993). Very recently, the deliberate practice framework has come into question as the single most authoritative explanation for how elite musicians refine and hone their skills at the highest levels of music training. This is based on assertions that deliberate practice may account for as little as 30% of the variance in expertise (Hambrick et al, 2013), as well as the many examples of exceptions to what have become known as the 10-year and 10,000-hours rules (Kaufman, 2013). Deliberate practice theory identified additional 'constraints'—on resources, motivation, and efforts—that limit people's capacities and abilities to undertake such vast quantities of practice, which to date are not adequately understood, and this research, in turn, has ramifications for musician's performance under pressure.

On a practical level, our work has attempted to extend this literature through recommendations of what musicians can do to increase the effectiveness of their learning and improve their performance. The deliberate practice approach has helped to pinpoint practice as the primary activity that improves performance, but an even more complete explanation for music education is required to understand how musicians can maximize their learning outcomes as they strive to develop musical expertise. Our research shows that *motivation* and *practice quality* are powerful predictors of performance. We have demonstrated that optimized motivation in music learners is associated with greater psychological fulfilment and valuing of music (Evans & McPherson, 2015; see also Renwick, 2008) and through our longitudinal research we have shown that adolescent musicians who were motivated by long-term goals had better long-term engagement and achievement over a period of more than 10 years (McPherson, Davidson & Faulkner, 2012). We have also found that musicians whose motivation was not optimal practiced more, but less effectively, than those whose motivation was optimal (Evans & Bonneville-Roussy, 2015). Our research on practice quality has found that better musicians use more sophisticated strategies (McPherson, 2005), are more confident about their abilities, practice in a more motivated and engaged manner (Miksza, 2011) and that all of these variables predict achievement above and beyond the quantity of practice they may have accumulated.

These findings in music education reflect what is found in other domains such as sports, health, education, personal relationships, and work environments. In these studies, constructs related to motivation and practice quality are consistently associated with improved performance, well-being, learning, and creativity (Ryan & Deci, 2000). One notable example is school education throughout the 20th century, which reformed from a teacher-centered model to a student-centered model that acknowledges the importance of a more supportive, facilitative environment, resulting in enormous improvements in motivation and educational outcomes compared with learning in competitive and controlling environments (Schunk, Meece & Pintrich, 2014). In research on school education, teachers who are autonomy supportive have students whose motivation and engagement is positively affected (Reeve & Halusic, 2009; Jang, Reeve, & Deci, 2010).

In this chapter, we explain research related to motivation and practice quality as a means of outlining our thoughts on how musicians are able to develop competence and acquire the skills necessary to perform under pressure. We start with motivation as viewed through the lens of self-determination theory (SDT). Since the 1970s, SDT has become a major approach to the study of human motivation. The theory rests on the assumption that humans naturally seek growth through interactions with their social environments. This growth is nourished by the fulfilment of three basic psychological needs (Deci & Ryan, 2000): the need to feel effective in interacting with the social environment (*competence*), the need to feel accepted by others (*relatedness*), and the need to feel that one's actions are aligned with one's interests and sense of self (*autonomy*). Practice quality is viewed through the lens of self-regulated learning (SRL; Zimmerman, 2000), a social-cognitive theoretical framework that is used to understand students' strategic behavior as they work through problems. Thus, self-determination theory helps us conceptualize how musicians of all ages are able to build the resilience and personal attributes necessary to cope with the demands of performing, particularly in stressful and demanding situations as often can occur in the competitive environments in music programs and in the professional lives of musicians.

Contemporary Psychological Perspectives on Motivation

Contemporary psychological views on motivation focus on multiple dimensions of people's experience in the social world–their thoughts, beliefs, and values, their interactions with the social environment, the quality of their relationships, their affective states, and so on. But this has not always been the case. The dominant paradigm for examining people's motivational states, particularly during the first half of the 20th century, focused almost exclusively on reward and punishment contingencies in the environment, and neglected to systematically study people's inner thoughts and representations. Known as behaviorism, this perspective conceived of learning as elaborate sequences of operant conditioning, administered by a socializer such as a teacher or parent (Schunk, Meece, & Pintrich, 2014).

More recent perspectives, rising from the advent of the 'cognitive revolution,' have rejected the idea that external events can comprehensively explain motivation. The cognitive perspective gave researchers a framework for studying constructs that are of more direct and obvious relevance to motivation–inner states such as beliefs and the way these might interact with the environment. One such inner state that has been shown to be important in motivation is the need to feel as though one's actions are self-endorsed, the belief that one's actions are aligned with one's interests, and the feeling that one's actions are engaged in freely and willingly. Collectively, this is known as autonomy (Ryan & Deci, 2000).

The following sections elaborate on research related to autonomy and show its importance in relation to motivation for instrumental music performance.

Motivation and Musical Development

Decades of research into the motivational underpinnings of self-regulated learning have been used to frame more sophisticated understandings of music learning (Renwick & Reeve, 2012; Evans, 2015; Austin, Renwick & McPherson, 2006). Much of this work is grounded in educational research focused on understanding learners' motivational beliefs and the learning strategies they recruit to achieve their goals. Music researchers have attempted to clarify the connection between music learning and the various motivational beliefs that influence the direction and maintenance of this self-regulatory process (Evans, 2015). One concept that has dominated motivation research is the distinction between engaging in an action or task because it is inherently interesting (intrinsic motivation) or because for some other reason—because it is useful in attaining a goal, because of pressure from another person, or because of the feeling of an ego boost (extrinsic motivation). SDT expands on this distinction with a more meaningful elaboration on extrinsic motivation.

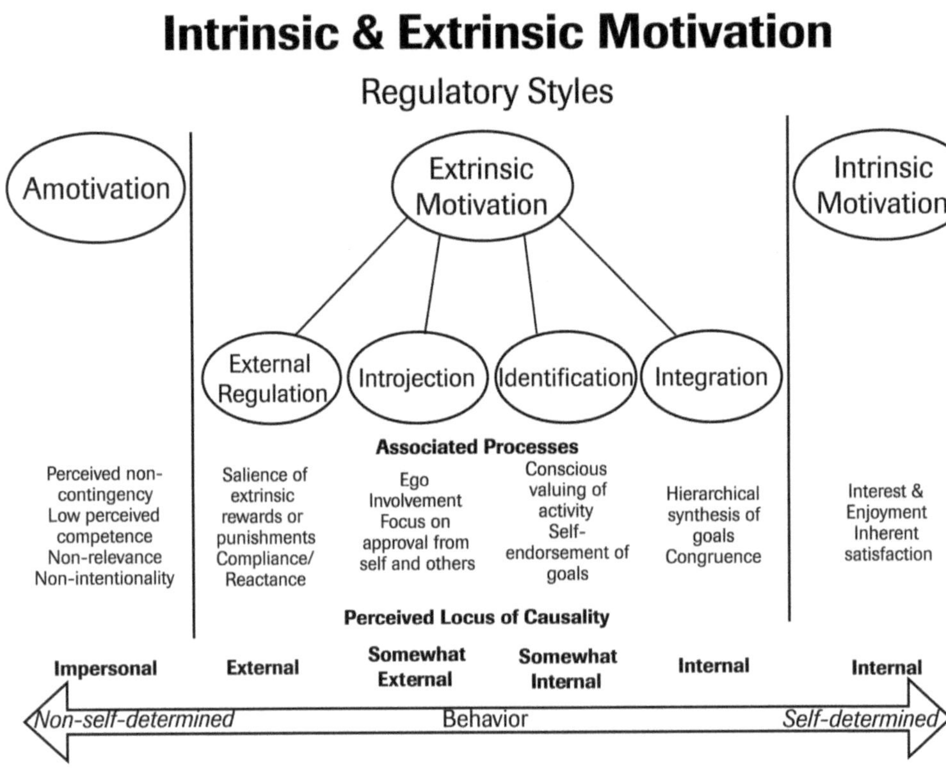

Figure 1: Self Determination Theory constructs. Adapted from Ryan & Deci, 2000, p. 61 and Renwick & Reeve, 2012, p. 14.

Young music learners commence learning musical instruments with a range of motivations that can be described along the extrinsic–intrinsic continuum. For example, some students may start learning an instrument at the suggestion of their parents, because their friends are learning, or because it is required by their school. Others might start for more internal reasons, such as wanting to play their favorite song on an instrument, or even being drawn to the sound of the instrument itself. Whatever the initial reason, motivation can be internalized depending on the circumstances of the social environment. Within SDT this integration process is a normal part of development (Ryan & Deci, 2000). Four regulatory styles on the continuum have been identified: (a) the purely extrinsically controlled *external* regulation where the importance of extrinsic rewards, punishment, or compliance dominate learning; (b) the *introjected* regulation controlled by such ego-related dynamics as guilt, self-esteem, and approval-seeking from oneself or others; (c) *identified* regulation, where there is a conscious valuing and self approval by the learner for undertaking the activity and one's motives are integrated into the self-schema; and (d) *integration* regulation where learners are able to internalize the reasons for their actions and assimilate them into their sense of self, even though these extrinsic motivation processes still rely on an outcome that is separate from the behavior (see Figure 1).

Young people's musical participation and engagement declines throughout their school years (McPherson & O'Neill, 2010; Wigfield et al., 1997). Researchers have attempted to explain the decline citing low levels of perceived musical competence (Wigfield et al., 1997) and intrinsic motivation (Hallam, 1998). Little research has been conducted, however, on the links between motivational beliefs and learning behaviors that might lead to a sense of competence and pleasure in music-making, which is quite remarkable given the widespread popular belief that learning an instrument teaches young people 'self-discipline.' Thus, self-determination theory is useful and a more sophisticated starting point for understanding how music learning can be optimized.

In formal learning environments, it is clear that learners may be motivated to develop competencies that are valued by others (e.g., teacher, parents, peers). According to SDT (Ryan & Deci, 2000), the foundation for facilitating internalization leading to the upper levels of extrinsic motivation and eventually intrinsic motivation rests on the ongoing need for learning to satisfy three psychological needs (*relatedness, competence, autonomy*) that allow learners to develop and function in a healthy and optimal way. First, learners need to experience a sense of belonging within the group or cultural practice in which they participate. Feelings of respect and being cared for within this environment by their teacher, parents, or others is essential if they are to willingly accept the values being imparted by others. Second, they need to feel competent. Learners do not usually persevere in learning difficult talks if they do not perceive the challenge as relevant and are able to receive constructive feedback during the process of mastering the skill or assigned task. Third, learners need to feel a sense of autonomy. This psychological need has received little previous attention in the music literature, even though it is paramount to understanding how music learning can be optimized. For this reason, we focus the following sections on this psychological need and the research related to this within other areas of learning and within music, especially with regard to how teachers can maximize their ability to impart skills, knowledge, and understanding through the use of support strategies that aid autonomy in music learners.

Providing Autonomy Support

As inferred in the previous paragraph, autonomy is central in young children's healthy development and especially so in young music learner's musical development where the social context plays an important role in development (Kupers, van Dijk, van Geert & McPherson, 2015; Kupers, van Dijk, McPherson & van Geert, 2014). Several studies on music lessons and other educational contexts have confirmed that teacher's autonomy support is associated with positive learning outcomes in learners. In a series of studies with certified Suzuki trained teachers, Kupers, van Dijk, van Geert, and McPherson (2015) examined how autonomy supportive environments support young music learners. Specifically, we sought to explore what types of teacher-student interactions best foster autonomy supportive learning and might best lead to higher engagement by the music learner. We proposed two dimensions for coding student states in string instrument lessons. The first dimension - *autonomy* - involved a measure of high versus low autonomy; the second - *engagement* - described the students' level of commitment as they were learning. The combination of level of autonomy and degree of engagement resulted in four quadrants of student autonomy expression as described below and in Figure 2.

Results of the study suggest the following:

1. *Autonomous engagement* (high autonomy and positively engaged in the learning task). The student would be taking on-task initiative by asking questions or making relevant on-task remarks.

2. *Resistance* (high autonomy and negatively engaged). The student would generally be actively resisting the task, by saying, for example, 'I

don't want to do this,' or by making off-task remarks such as 'I'm going to the playground later.'

3. *Mimicry* (low autonomy and positively engaged). The student would be 'going with the flow,' doing what was expected by the teacher and not taking any personal initiative.

4. *Absence* (low autonomy and negatively engaged). The student would not be engaged with the task, would not be responding verbally (i.e., not answering teacher questions), but would also not be actively resisting it (see Kupers, van Dijk, van Geert & McPherson, 2015, pp. 340-341).

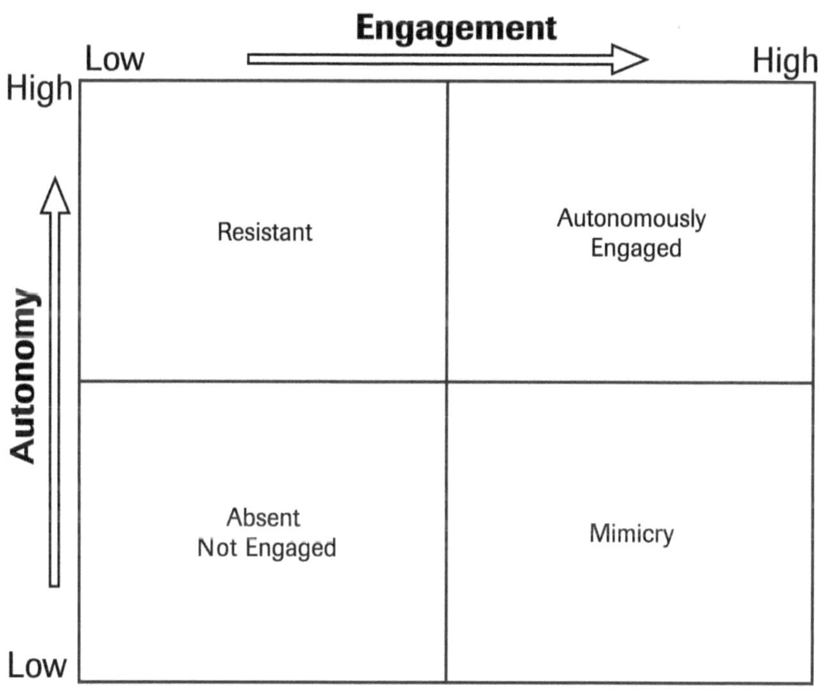

Figure 2. Four quadrants of student states of autonomy (see Kupers, van Dijk, van Geert & McPherson, 2015, p. 341.)

First, from the analyses, we observed large differences between the dyadic transactions of students with a higher overall need for autonomy, compared to students with a low need for autonomy. Instead of the teacher being the 'locus of control' in terms of autonomy support, in these situations, autonomy was more 'negotiated' between teacher and student. At other times, the autonomy support and autonomy expression levels of the teacher and music learner were well coordinated, in contrast to other moments where there were large discrepancies.

These results led us to question the function of dyadic synchrony in autonomy levels when learning to play an instrument. Relevant literature dealing with synchrony in parent-infant interactions (e.g., Harris & Waugh, 2002) shows that more adaptation or synchrony is related to healthy developmental outcomes. As we have shown in our research, any friction in synchronization between verbal interactions (such as when a teacher offers autonomy support at a higher level than the student's current level of autonomy) will help propel and maintain momentum in the learning process. The results of our work suggest a number of implications. One is that our work outlines strategies that teachers can use to foster their young music learners' autonomy. At another level, our work also shows that there is no 'one size fits all' with regard to autonomy development and the techniques teachers use to provide autonomy support. What works for one student may not necessarily work for another music learner. What now needs to be studied in more detail, however, is the moment to moment synchronicity in teacher-student interactions. Obviously friction leading to negative consequences in students is something that needs to be carefully avoided, whereas a moderate amount of asynchrony in the lesson may well have a role in challenging and stimulating students. Much more research is needed to explore and understand these mechanisms.

Suggestions for Improving Teaching Practice

Supporting students' autonomy involves supporting and reinforcing a constellation of behaviors that support students' feelings of choice, so that their experience of autonomy derives from their sense of being in control of a task, as opposed to feeling controlled by the task. Autonomy support therefore describes the processes teachers adopt during their teaching to facilitate, encourage, and develop their students' innermost motivational resources (Reeve & Halusic, 2009). The opposite of *autonomy support* is *behavioral control*. Behavioral control occurs when the behavior of a teacher acts to suppress or coerce students into behaving in a particular manner. Students who have autonomy-supportive teachers are more likely to thrive in their learning, engagement, and well-being, while students who have controlling teachers are likely to become disengaged and less motivated (Reeve & Halusic, 2009).

Citing the extensive body of research on this topic, Renwick and Reeve (2012) have identified five ways teachers can support autonomy in music:

> Spark students' initial engagement in learning activities by nurturing their inner motivational resources with comments such as "Here is a challenge; here is a piece of music that will challenge your skills."
>
> Provide a steady stream of rationales to explain the *why* behind any potentially uninteresting endeavor by using comments such as: "Okay, playing the same piece of music over and over tends to get boring, but the reason for the repetition is to make gradual little refinements each time until, in the end, you have mastered the piece of music."
>
> Rely on non-controlling and informational language when they communicate requirements, comment on student progress, ask students to take responsibility for their learning, and address motivational and behavioral problems through the use of comments such as: "I don't hear much progress; do you know why that might be?"
>
> Display patience for self-paced learning to occur by providing sufficient time in lessons to listen to students work in their own way, and when they seem stuck, postpone advice until it is understood what the students are trying to accomplish.
>
> Acknowledge and accept students' expressions of resistance and negative emotions when they complain that the task is too boring or too hard, with comments such as: "Yes, this is hard; it can make anyone feel frustrated. What could we do differently so it didn't seem so boring or difficult; any suggestions?" (see Renwick & Reeve, 2012, p. 149)

The one-to-one master-apprentice style of music lesson is often characterized by approaches that too often undermine student autonomy and motivation. In many situations, teachers are highly controlling in the choice of repertoire they allow students to learn, and the ways in which they allow their students to master or interpret the repertoire they are assigned to learn. Accordingly, interventions to avoid the natural inclination of master musicians to be controlling when giving lessons involve encouraging them to adopt a much more autonomy supportive motivational style that finds ways of empowering students through efforts by the teacher to take into account and value the student's perspective (Evans, 2015; Renwick & Reeve, 2012). Highly skilled musicians can achieve this by inviting and welcoming their students to express their thoughts, feelings, and actions within the lesson so that their instruction is not predetermined and unilateral but negotiated at each step of the learning process (Renwick & Reeve, 2012, pp. 156-157).

The five autonomy supportive instructional behaviors explained above can be used to help teachers rethink how they support their music learners (Renwick & Reeve 2012). First, instead of telling students what to do and giving them little choice in how they learn, music teachers should attempt instead to nurture students' inner motivational resources in ways that allow them to set their own goals, and discover and find for themselves what they find most interesting and valuable. Second, when engaging students in uninteresting activities such as learning technical exercises, music teachers should carefully contextualize the reasons, the need, and the benefits of the task so that their students understand the worth of the activity in terms of improving their ability. Third, by adopting non-controlling comments during lessons, students can be encouraged to optimize their own learning strategies and reflect on the ways that they feel are most beneficial for them to reach their personal best. Fourth, being patient and allowing time for students to self-pace their own learning, set their own goals, diagnose problems for themselves, and reflect on their own progress encourages learners to take control of their own learning as they come to realize how difficult, challenging tasks take time to master. Finally, to encourage persistence and to overcome obstacles, it is important for music teachers to acknowledge and manage students' frustrations and negative emotions as a normal part of any form of high-level skill development.

Optimizing Music Instruction

A key principle of SDT is that systems of teaching and learning must ensure that a sense of relatedness, competence, and autonomy are supported as a means of motivating young musicians to employ the self-regulatory skills that likely lead to high level engagement and ultimately advanced levels of musical competence. At the simplest level, music teachers can be advised to focus on encouraging effort management, self-monitoring, and adaptive strategy use, instead of emphasizing time spent practicing. For instance, in some music programs, children are given notebooks in which they are required to state how many minutes of practice they have completed each day. This approach might simply reinforce a focus on the passage of time (McPherson & Renwick, 2001), rather than the mastery of the musical skill or task at hand. An approach that might encourage more autonomy and engagement would be to ask students to use a practice journal to record and reflect on the strategies they choose in order to maintain effort, to monitor accuracy, correct errors, and develop their own interpretation of the piece. Such efforts to develop greater engagement have been found by McPherson, Davidson and Faulkner (2012) to increase practice efficiency.

Extending the above, we might examine some of the various models of leadership, teaching, and parenting styles. As an example, Dinham and Scott (2008) developed Baumrind's (1991) earlier work on effective parenting to produce four prototypes of leadership or "teachership." As can be seen in Figure 3, these four prototypes are categorized along two dimensions. The first is the degree to which the teacher is *responsive* to being able to deal emotionally and intellectually to students' needs or those of the teaching environment. The second concerns *demandingness*, or the way the teacher might insist on adopting a certain way of working, irrespective of how students might feel or wish to work themselves.

The four teaching styles depicted in Figure 3 can be defined as follows:

Uninvolved Teachers are low in 'responsiveness' and 'demandingness.' Under such forms of tuition students would be simply left to their own devices with their teacher having little if any positive impact on their education. Obviously, this style of teaching would rarely be seen in schools.

Authoritarian Teachers are high in demandingness and low on responsiveness, meaning that they expect compliance from their students and are unwilling to negotiate or consult. Authoritarian Teachers also tend to rely heavily on rewards and rules, and whilst achievements may be high, there is often an ethos of fear or coercion that can be stressful for the student.

Permissive Teachers are high in 'responsiveness' and low in 'demandingness.' They may allow a considerable amount of freedom, have high interpersonal skills, and value consensus. Cultures or systems led by them are often happy places but there is a risk of underachievement and lack of direction especially for those individuals with low levels of self-regulation. Other highly motivated individuals may excel.

Finally, *Authoritative Teachers* have high levels of both 'responsiveness' and 'demandingness.' According to Dinham and Scott (2008), such teachers are concerned with all of their students experiencing success and achievement.

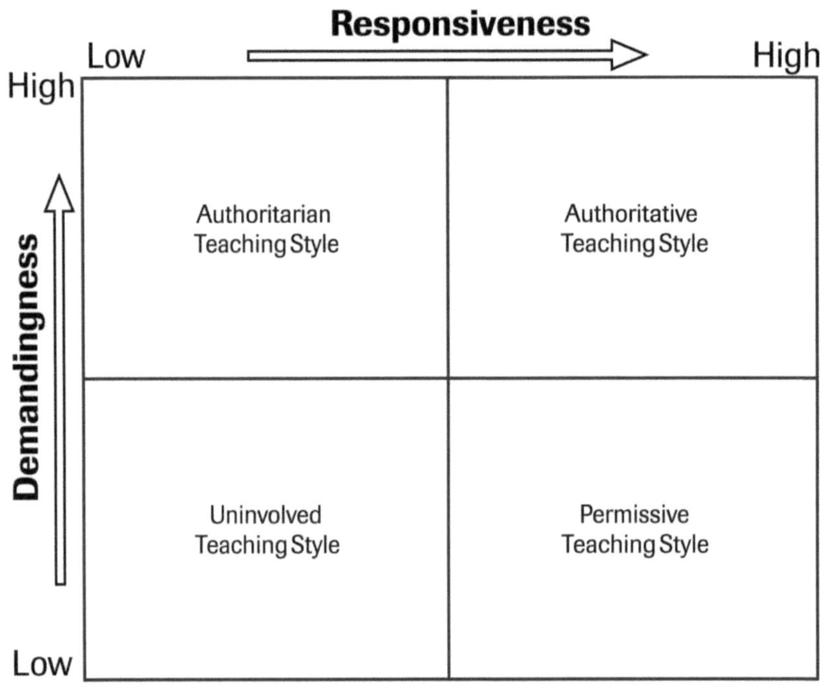

Figure 3: Four prototypes of teaching styles (adapted from Dinham & Scott, 2008, after Baumrind, 1991.)

Self-Regulated Learning and Practice Quality

Complementing research in motivation is research on the things that learners do to set goals, monitor their progress towards them, evaluate the effectiveness of their strategies, and reflect on their learning. Collectively, these processes are known as *self-regulated learning* (SRL; McPherson & Renwick, 2011; McPherson & Zimmerman, 2011). According to the SRL framework, effective learners engage in three processes in a self-regulated learning cycle:

Forethought: Setting appropriate goals and identifying effective strategies that can be used to achieve them (e.g., setting a goal of effectively playing through an 8-bar section of a new piece of music that has been identified as important; using a strategy of first slowing the tempo, then gradually increasing it as performance is accomplished).

Performance: Engaging in tasks mindfully, monitoring performance during a task (e.g., observing that the playing is fluent at the current tempo and increasing the tempo accordingly; identifying that a particular fingering pattern is not proving to be effective and trying two alternatives).

Reflection: Attributing performance to causes (e.g., I didn't practice effectively today because the strategies I identified weren't appropriate) and preparing for the following iteration of the cycle (e.g., learning that 8-bar section was easy today; perhaps next time I can be more ambitious about what I fit into a practice session).

In McPherson's (2005) study of children learning music, a series of tasks was presented to the children. One of these was a test of musical memory—a short notated melody was presented briefly, then the children were asked to play the melody from memory. The children reported using a range of strategies to deal with the problem. Some reported relatively poor and unsophisticated strategies: "I remember the names of the notes;" "I try to take a mental photograph of the music so I can remember it when it is no longer there." Others used more sophisticated strategies: "I sing the melody in my head and imagine playing through it on my instrument." In this study, these self-regulatory strategies were a more consistent predictor of performance on measures of music performance than practice quantity. Given the dominance of practice quantity (or accumulation of hours of deliberate practice) in the literature, the finding that practice quality may have a greater impact speaks to the potential for self-regulated learning as a framework for understanding music learning.

Self-regulated learning also appears to be a useful way to study university students' motivations for music learning. Miksza (2011) examined self-regulated learning in brass and woodwind players. His observations reflected the deliberate practice literature, showing that most university music students in the study were practicing in a way that would be consistent with a definition of deliberate practice. However, within this range, those who had used more sophisticated strategies had higher performance achievement scores. This research contributes to understandings about how music-specific self-regulated learning strategies may improve performance. More recent experimental research by Miksza (2015) suggests that self-regulated learning strategies can be taught relatively easily, and that students who are taught them practice more musically (incorporating more expressive elements and focusing on performance aspects such as phrasing), than others (who practiced in a basic way by focusing only on the sequence of notes and rhythms).

Integrating Motivation and Practice Quality

Having reviewed the areas of motivation (using SDT as a framework) and practice quality (using SRL), we are now in a position to examine the extent to which these perspectives may provide complementary explanation for ongoing achievement in music, as well as other outcomes of interest, such as well-being, music performance ability, and a long-term commitment to engagement in music. Some findings in the literature lead us to believe that these perspectives may indeed be complementary and provide a fuller explanation for music performance than perspectives such as the deliberate practice framework, which focuses almost exclusively on the accumulation of hours of deliberate practice. One finding comes from one of our case studies with a musician who, when autonomously selecting a piece of music that she was interested in, showed vastly superior practice strategies compared with previously, when she had been playing repertoire selected for her by her teacher (Renwick & McPherson, 2002). In this case study, the student was observed to be autonomous in her motivation, and subsequently, she spontaneously adopted quality self-regulated learning techniques. Thus, we hypothesize a causal link between motivation and practice quality.

Another more recent finding comes from a study of university music learners (Evans & Boneville-Roussy, 2015). In a survey of learners from a number of institutions, motivation was profiled using self-determination theory. Participants reported on the number of hours they had engaged in practice, the productivity of their practice, and their preference for expanding their learning by choosing challenging tasks, rather than tasks that were easy and within their ability. Strong links were found between students' psychological needs and their motivation, and, in turn, their motivation and measures of practice. Interestingly, motivation predicted practice quality more powerfully than practice quantity.

Therefore, there is some evidence to suggest that motivation and practice quality may be complementary approaches that more fully explain music performance achievement.

Conclusion

Motivation and practice quality are two key aspects of music learning that seem relevant from childhood to adolescence and adulthood, and in skill levels ranging from novice to expert. Motivation is particularly relevant to teaching, as our research has demonstrated that the quality of the relationship makes a difference to learning. The way in which teacher's level of autonomy support and the student's state of autonomy are 'matched' or 'mismatched' from moment to moment can be linked to long-term-outcomes. We suggest that there are more and less optimal styles for music teachers to support students' sense of autonomy, to ultimately foster their intrinsic motivation. However, perhaps even more important than having a general 'style' of teaching, is the teachers' ability to be able to show flexibility in adapting their level of responsiveness and demandingness to the current needs of individual students.

In terms of practice quality, it is important for teachers to understand that most of the work done by students to improve is done on their own. Therefore, it is important to equip them with the tools and skills necessary to practice effectively. In our view, this is best achieved by encouraging them to develop self-regulated practice strategies, by encouraging them to become aware of the relevant forethought, performance, and reflection processes necessary for engagement in productive learning.

On an individual and personal level, self-determination theory allows us to also conceptualize how external control of any kind can undermine a musician's motivation to achieve at the highest level, particularly in stressful situations where the musician is required to perform under pressure. This theory helps us understand how musicians can feel in control of their musical activities, or, in contrast, feel overwhelmed by certain stressful situations. Often the root cause of such feelings is how musicians view the activity: in short, whether they feel in control of the musical situation or, alternatively, whether they feel that the musical situation controls them. Understanding these mechanisms provides an important means for helping musicians who are struggling and feeling under pressure to build their confidence in ways that are much more deeply important to them.

Each of the four authors of this chapter has conducted research using the self-determination and self-regulated learning frameworks. We hope that the brief summary of our work in this chapter will encourage other researchers to expand knowledge in this area, as collectively, we search for even better ways to optimize music students' potentials, and grow understandings of how all musicians can develop the resilience to cope with the enormous challenges required to master music performance.

Key Additional Resources

To further understandings of the material covered in this chapter, readers are encouraged to access the following dissertations from the second, third and fourth authors.

Evans, P. (2009). *Psychological needs and social-cognitive influences on participation in music activities* (Order No. 3362780). Available from ProQuest Central; ProQuest Dissertations & Theses Global. (304895600).

Kupers, E. (2014). *Socially situated learning in individual music lessons.* Unpublished doctoral thesis, University of Groningen. Retrieved from https://www.rug.nl/research/portal/publications/pub(08de217f-c6f5-4b95-9fee-f2ce592ab501).html

Renwick, J. M. (2008). *Because I love playing my instrument: Young musicians' internalised motivation and self-regulated practising behaviour.* Unpublished doctoral thesis, University of New South Wales. Retrieved from http://handle.unsw.edu.au/1959.4/36701

References

Austin, J. R., Renwick, J., & McPherson, G. E. (2006). Developing motivation. In G. E. McPherson (Ed.), *The child as musician: A handbook of musical development* (pp. 213–238). Oxford: Oxford University Press.

Baumrind, D. (1991). The influence of parenting style on adolescent competence and substance use. *The Journal of Early Adolescence, 11*, 56-95.

Deci, E. L., & Ryan, R. M. (2000). The “what” and “why” of goal pursuits: Human needs and the self-determination of behavior. *Psychological Inquiry, 11*, 227–268. doi:10.1207/S15327965PLI1104_01

Dinham, S., & Scott, C. (2008). Responsive, demanding leadership. *Management Today, 43*, 32-35

Ericsson, K. A., Krampe, R. T., & Tesch-Römer, C. (1993). The role of deliberate practice in the acquisition of expert performance. *Psychological Review*, 100, 363–406.

Evans, P. (2015). Self-determination theory: An approach to motivation in music education. *Musicae Scientiae*, 19, 65–83. doi:10.1177/1029864914568044

Evans, P., & Bonneville-Roussy, A. (2015). Self-determined motivation for practice in university music students. *Psychology of Music.* doi:10.1177/0305735615610926

Evans, P., & McPherson, G. E. (2015). Identity and practice: The motivational benefits of a long-term musical identity. *Psychology of Music, 43*, 407–422. doi:10.1177/0305735613514471

Hallam, S. (1998). The predictors of achievement and dropout in instrumental tuition. *Psychology of Music, 26*, 116–132.

Hambrick, D. Z., Oswald, F. L., Altmann, E. M., Meinz, E. J., Gobet, F., & Campitelli, G. (2013). Deliberate practice: Is that all it takes to become an expert? *Intelligence*. doi: 10.1016/j.intell.2014.04.001

Harrist, A. W., & Waugh, R. M. (2002). Dyadic synchrony: Its structure and function in children’s development. *Developmental Review, 22*, 555-592.

Jang, H., Reeve, J., & Deci, E. L. (2010). Engaging students in learning activities: It is not autonomy support or structure, but autonomy support and structure. *Journal of Educational Psychology, 102*, 588–600. doi:10.1037/a0019682

Kaufman, S. B. (Ed., 2013). *The complexity of greatness: Beyond talent or practice.* Oxford, England: Oxford University Press.

Kupers, E., van Dijk, M., McPherson, G. E., & van Geert, P. (2014). A dynamic model of skill acquisition and self-determination in instrumental music lessons. *Musicæ* Scientiæ, *18*, 17-34.

Kupers, E., van Dijk, M., van Geert, P., & McPherson, G. E. (2015). A mixed-methods approach to studying co-regulation of student autonomy through teacher-student interactions in music lessons. *Psychology of Music, 43*, 333-358.

McPherson, G. E. (2005). From child to musician: Skill development during the beginning stages of learning an instrument. *Psychology of Music, 33*, 5–35.

McPherson, G. E., & O’Neill, S. A. (2010). Students’ motivation to study music as compared to other school subjects: A comparison of eight countries. *Research Studies in Music Education, 32*, 101–137. doi: 10.1177/1321103x10384202

McPherson, G. E., & Renwick, J. M. (2001). A longitudinal study of self-regulation in children’s musical practice. *Music Education Research, 3*, 169–186.

McPherson, G. E., & Renwick, J. M. (2011). Self-regulation and mastery of musical skills. In B. J. Zimmerman & D. H. Schunk (Eds.), *Handbook of self-regulation of learning and performance* (pp. 234–248). New York: Routledge.

McPherson, G. E., & Zimmerman, B. J. (2011). Self-regulation of musical learning: A social cognitive perspective on developing performance skills (pp. 130-175). In R. Colwell & P. Webster (Eds.), *MENC handbook of research on music learning* (vol. 2). New York: Oxford University Press.
McPherson, G. E., Davidson, J. W., & Faulkner, R. (2012). *Music in our lives: Rethinking musical ability, development and identity.* Oxford, England: Oxford University Press.
Miksza, P. (2011). The development of a measure of self-regulated practice behavior for beginning and intermediate instrumental music students. *Journal of Research in Music Education, 59*, 321-338. doi: 10.1177/0022429411414717
Miksza, P. (2015). The effect of self-regulation instruction on the performance achievement, musical self-efficacy, and practicing of advanced wind players. *Psychology of Music, 43*, 219–243. doi:10.1177/0305735613500832
Reeve, J., & Halusic, M. (2009). How K-12 teachers can put self-determination theory principles into practice. *Theory and Research in Education, 7*, 145–154. doi:10.1177/1477878509104319
Renwick, J. M. (2008). *Because I love playing my instrument: Young musicians' internalised motivation and self-regulated practising behaviour.* Unpublished doctoral thesis, University of New South Wales. Retrieved from http://handle.unsw.edu.au/1959.4/36701
Renwick, J. M., & McPherson, G. E. (2002). Interest and choice: Student-selected repertoire and its effect on practising behaviour. *British Journal of Music Education, 19*, 173–188. doi: 10.1017/S0265051702000256
Renwick, J. M., & Reeve, J. (2012). Supporting motivation in music education. In G. E. McPherson & G. Welch (Eds.), *The Oxford handbook of music education.* New York: Oxford University Press.
Ryan, R. M., & Deci, E. L. (2000). Self-determination theory and the facilitation of intrinsic motivation, social development, and well-being. *American Psychologist, 55*, 68–78.
Schunk, D. H., Meece, J. L., & Pintrich, P. R. (2014). *Motivation in education: Theory, research, and applications* (4th ed.). Upper Saddle River, New Jersey: Pearson.
Wigfield, A., Eccles, J. S., Yoon, K. S., Harold, R. D., Arbreton, A. J. A., Freedman-Doan, C., et al. (1997). Changes in children's competence beliefs and subjective task values across the elementary school years: A 3-year study. *Journal of Educational Psychology, 89*, 451–469.
Zimmerman, B. J. (2000). Attaining self-regulation: A social cognitive perspective. In M. Boekaerts, P. R. Pintrich & M. Zeidner (Eds.), *Handbook of self-regulation* (pp. 13–39). San Diego, CA: Academic Press.

Under Pressure: Music-Related Interventions in High-Performance Domains

The title of this chapter echoes that of a well-known song by supergroup Queen – *Under Pressure*. Anyone who is engaged in a high-performance domain, such as a sporting contest, a musical recital, or a marketing pitch to a high-powered executive board, will know only too well what it feels like to be under pressure. Nonetheless, dealing effectively with pressure is part and parcel of such high-performance domains; when you go for a walk in the rain you will expect to get a little wet, when you enter a high-performance domain you will expect to feel some pressure. Seen in this light, pressure is *perhaps* a sign of success.

For certain individuals, the pressure will serve to raise their performance to extraordinarily high levels while for others, the spotlight of public scrutiny can be overwhelming and they crack like eggs in an omelet (Clark, Lisboa, & Williamon, 2014; Neil, Wilson, Mellalieu, Hanton, & Taylor, 2012). When they disintegrate, a phenomenon that psychologists often refer to as 'choking,' it is often due to an excessive focus on the execution of key skills rather than allowing such skills to flow automatically. The awareness of one's own anxiety levels prompts a debilitating effort to consciously and falteringly implement skills that have long since become automatic and effortless. This distraction triggers a meltdown in which action is choked by endless negative speculations and other mental flight.

The purpose of this chapter is to explore the structured and systematic use of music as a means to optimize the mindset and outcomes for those involved in high-performance domains. The first section of the chapter deals with key concepts and some underlying scientific principles. The second section presents a conceptual framework that is intended to underpin the music-related applications that are presented in the third section (e.g., using music to prepare mentally for a concert performance). The content of the chapter is likely to be of particular interest to athletes, performing artists, business people, coaches, performance psychologists, financial traders, armed-forces personnel, surgeons, and teachers.

The Special Role of Music

From the dawn of civilization, ancient cultures have sought to combine sounds in such a way that they influence the human psyche. With the evolution of such cultures, primitive forms of music morphed into ever-more artistically pleasing arrangements. Composers and musicians became increasingly adept in using musical works to alter the emotional state of their listenership. For example, the preeminent classical composer Ludwig van Beethoven (1770-1827) was famed for his ability to stir human emotions through the rich texture and intricate use of sonic tension and resolution in his works. Seemingly, there is a piece of music for every civic occasion as well as for almost every ritual and function of our private lives. Accordingly, through the ages, music became intertwined with a diverse range of human activities that included the worship of deities, soothing of tearful infants at bedtime, priming of soldiers facing battle, celebration of rites of passage, and synchronization of manual labor.

Musical compositions entail the organization of four primary elements: *melody*, *harmony*, *rhythm*, and *dynamics*. Melody is the part of a musical composition that is often the highest in pitch and colloquially known to as 'the tune.' It is the part that you might catch yourself humming or whistling along to. Harmony describes the combination of multiple notes and often serves to shape the emotional impact of the music. If you can imagine a barbershop quartet in full flow, that would be an example of four-part vocal harmony. Rhythm has to do with the tempo or speed of a piece of music—normally measured in beats per minute (bpm)—as well as the way that the music is accented over time. It is these accents that enable us to clearly distinguish between a reggae skank[1], a salsa montuno[2], and a Viennese waltz[3].

Rhythm is also the musical ingredient that prompts a physical response in the listener, such as the desire to tap the toes or dance. Dynamics concern the energy transmitted by musicians through their touch or breath to influence how loud their instrument sounds, particularly how this changes over the course of a piece. As we will discover later in this chapter, the emphasis that is placed on such musical qualities can have a direct influence on an individual's mindset as they ready themselves for a superior performance.

1 A skank is a West Indian colloquialism representing the beat that characterizes reggae music. This entails a strong emphasis on the second and fourth beat of the bar.

2 Montuno is a name given to the style of piano playing that typifies salsa music. It features a catchy, syncopated vamp that is repeated throughout the piece.

3 A Viennese waltz is an early genre of ballroom dance characterized by three beats to the bar with a strong emphasis on the first beat of the bar - **one**, two, three, **one**, two, three, etc.

Music Applications

Music can serve three primary functions in performance domains. First, it can be used *pre-performance* to manipulate a performer's mindset prior to the activity itself. This application is salient in terms of engendering the appropriate type of imagery, cognitions, and the optimal level of activation in the performer. By this we mean that they see the 'right' pictures in their mind, have positive and not self-defeatist thoughts, and are suitably aroused for the task at hand. Generally speaking, fast, loud music has a stimulative effect so it can be used as part of a psych-up routine whereas soft, slow music has a sedative effect, which means that it can be used as a relaxant. A performance that requires deep concentration and fine motor control, such as playing Beethoven's *Moonlight Sonata* on the piano, would require far lower levels of arousal than a gross motor performance, such as a bench press in the sport of powerlifting. In addition, well-selected pre-performance music will prompt emotions in the performer that they wish to take into their activity. For example, athletes in combat sports such as wrestling and boxing often choose highly aggressive music to accompany their entry into the ring (see box below).

Wrestling With the Choice of Music

In US professional wrestling, the much-feared Steve Austin was well known for parading into the ring while accompanied by the chilling instrumental track *Stone Cold Steve Austin*. This heavy metal piece was composed especially for the Texan and served to embody his implacability in the field of battle. Austin's long-time rival Hulk Hogan chose *Real American* as his walk-on music. This brash, guitar-led anthem is said to represent America and real American heroes, as is evident in the lyrics of the refrain, which allude to a fight to maintain the ideals that Americans hold dear. Hulk's public persona as the 'small-town guy with a big heart' that the crowd should rally behind was entirely in keeping with his choice of music. For both of these heavyweight contenders, the primary purpose of the music was to say "Hey people, here I am ... I am a mean dude, pumped up to the max, and ready to win big!"

The second primary application of music entails its *in-performance* use. The music in this case accompanies a performance or training session. With this type of application, we exclude the influence of a musical work on an artist *while* they are performing, which is not a focus of the present chapter. With such in-performance use, there are two subtypes of application. The *synchronous* application of music is typified by the use of the rhythmic aspects of music as a type of metronome that regulates movement patterns. For example, the rowers on a dragon boat employ a drummer to coordinate their collective efforts in time and space.

The *asynchronous* use of music, as the term implies, entails the absence of *conscious* synchronization. It is when music is playing in the background during an activity to enhance the performance environment or lift the mood of the performer(s). A good example would be international beach volleyball championships that have loud asynchronous music blaring over the public address system. The astute reader will notice a potential oxymoron here: Although asynchronous music is often referred to as 'background music' owing to the lack of synchronization, it *can* be very loud!

The third primary application of music is its *post-performance* use. This is when it functions in a recuperative role and aids recovery from injury, competition, or extreme physical exertion (Terry & Karageorghis, 2011). Selections with a tempo in the range 60-90 bpm are recommended when using music post-performance, as this is close to the average resting heart rate. In addition, the emotional qualities of the music should be neutral or relaxing and played on soothing instruments such as strings, clarinet, or piano in order to facilitate recuperative feeling states (Karageorghis & Terry, 2011).

How Music Affects Us

The literature pertaining to performance-related domains has focused on the *psychological*, *psychophysical*, *psychophysiological*, and *ergogenic* effects of music. The psychological effects relate to the influence of a musical composition on mood, emotion, affect (feelings of pleasure/displeasure), thought processes, and behavior. Psychophysical effects entail the psychological perception of one's physical state and therefore such effects are often assessed in the sports domain using Gunner Borg's Rating of Perceived Exertion (RPE) scales (see e.g., Hutchinson, Karageorghis & Jones, 2015; Lim, Karageorghis, Romer & Bishop, 2014; Stork, Kwan, Gibala, & Martin Ginis, 2014). The psychophysiological effects have to do with how music influences physiological functioning, that is whether music listening changes heart rate, breathing rate, blood pressure, etc. Finally, music has an ergogenic effect when it enhances a performer's work output or causes higher than expected power output, endurance, or productivity. Figure 1 provides a visualization of elite triathletes' affective responses during a progressive running task on a treadmill. The athletes' feelings of pleasure were far more positive in the presence of motivational music when compared to *oudeterous* (music that is neither motivating nor demotivating), and a no music control.

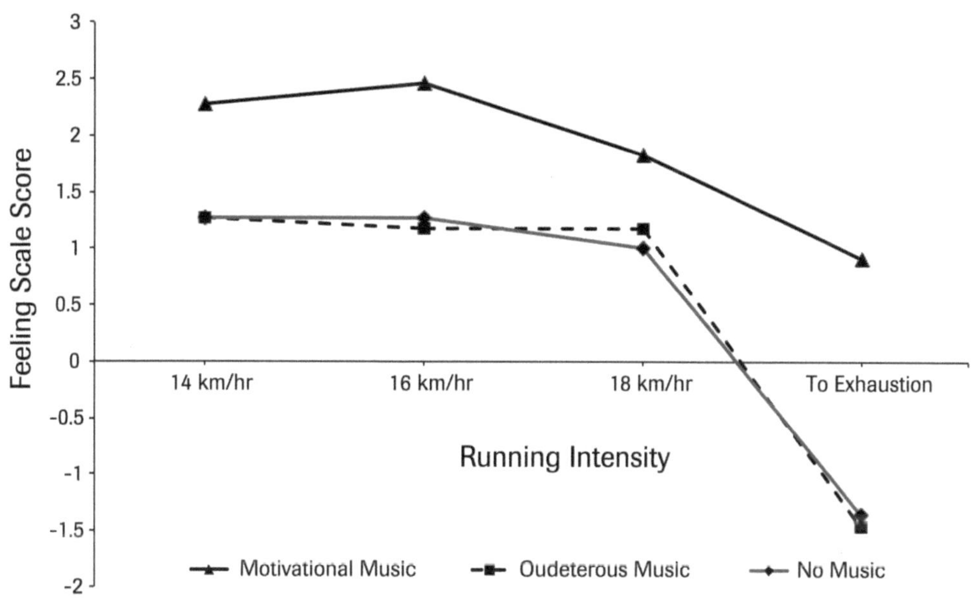

Figure 1: Feeling Scale scores of elite triathletes as the run to exhaustion progressed.[4]

Adapted from Journal of Science and Medicine in Sport, 15(1), P. C. Terry, C. I. Karageorghis, A. Mecozzi Saha, and S. D'Auria, Effects of synchronous music on treadmill running among elite triathletes, 52-57, © 2012, with permission from Elsevier.

An Underlying Conceptual Framework

I, Karageorghis, recently published a theory to predict the effects of music in the exercise and sport domain and an accompanying model that is *heuristic* in nature (2015). By heuristic I meant that the model is *instructive* and provides a holistic representation of relationships identified within the literature, as opposed to a mechanistic or factorial model that offers a series of explicit predictions. Herein, the theory and model are adapted

4 Note: There was a brief recovery period in between each of the four running intensities.

to a broader range of high-performance domains and presented in brief here as a conceptual framework to provide a backdrop for the range of music applications that follow in the second half of this chapter. It should be noted that this framework does not cover the influence of a piece of music on artists *while* they are performing that piece or to that piece. Rather, from the perspective of the performing artist, the theory pertains to how s/he might use music as part of her/his pre-performance preparation or post-performance recuperation. Figure 2 provides a visualization of the conceptual framework that describes the way in which variables pertaining to music, the listener, and the listening situation interact to determine a range of consequences in high-performance domains.

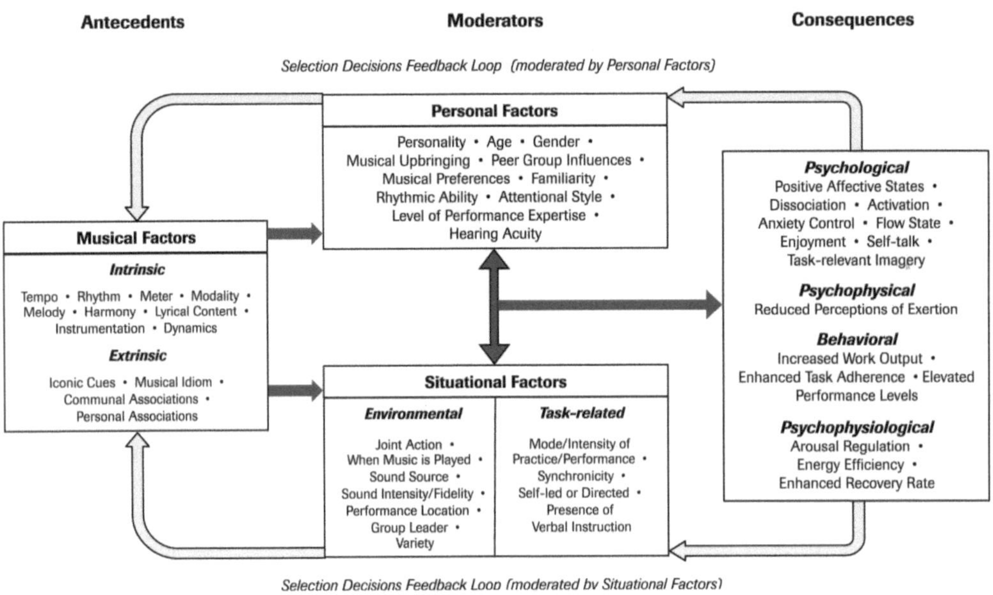

Figure 2: A conceptual framework of the antecedents, moderators, and consequences of music use in high-performance domains.

Adapted from Sport and Exercise Psychology (2nd ed., p. 301), by A. M. Lane (Ed.), 2015, London, UK: Routledge. © 2015, with permission from Routledge.

Musical Factors

Musical attributes are referred to as "antecedents" in Figure 2 because they precede or cause the responses to music. Think of them as the beginning of a type of 'chain reaction' that takes place when a performer is influenced by a musical stimulus. These attributes are divided into *intrinsic* and *extrinsic* categories: Intrinsic qualities relate to the constituent parts of music (e.g., melodic and harmonic features, vocalizations, syncopation, etc.), whereas extrinsic qualities pertain to how music is interpreted by the listener and the associations of the pieces (e.g., how it reminds the listener of other things).

Among the intrinsic properties of music, lyrics play a salient role through providing affirmations or directing attention toward a particular image, thought pattern, or skill (e.g., a high trapeze artist using the track *I Believe I Can Fly* by R Kelly). The influence of the meaning of the lyrics (i.e., their semantic detail) depends not only on the way it is received by the performer but also the relevance of the information to the task at hand. A particular lyric can be highly emotive for a given performer, and even imbue them with feelings of invincibility, but have no influence whatsoever on another (Juslin, 2013a; Karageorghis et al., 2013; Sanchez, Moss, Twist, & Karageorghis, 2014).

Among music's extrinsic properties, *iconic cues* concern how structural elements of a musical work relate to the tone of certain emotions; for example, fast/loud music may sound 'lively' because there are commonalities with energy and excitement (North & Hargreaves, 2008). Given that such cues are grounded in the structure of music, the same music should hold similar 'iconic meaning' for people of different cultures (Fritz et al., 2009).

The conceptual framework distinguishes communal associations from personal associations (Juslin, 2013b). The former are often sparked and galvanized by the mass media. Thus they are likely to influence large sections of the population, albeit for different reasons. For example, the use of saxophonist Paul Desmond's jazz composition *Take Five* to promote the sale of Cadbury's *Twirl* chocolate bars on TV might well compel younger viewers to purchase the sweet snack when they 'take five.' In contrast, older viewers might associate the music in the advertisement with '60s chic' and Desmond's iconoclastic approach to regular jazz meter[5]. Given that the present-day youngsters did not grow up listening to the piece, they are more likely to associate the piece with the chocolate bar whereas the same association for older viewers is less likely.

5 Regular jazz meter entails the use of four beats in a bar (quadruple time) whereas Desmond's *Take Five* was characterized by five beats in the bar (quintuple time).

On the other hand, personal associations pertain to our individual experiences of music. For example, the well-documented 'our song' phenomenon, which concerns how a particular piece of music can hold special meaning for lovers, reminds them of the time that they first started dating (North & Hargreaves, 2008). These associations are cultivated at a personal level but may involve the mass media; hence, there is some overlap with communal associations.

Personal and Situational Factors: The Moderators

Moderators are factors that influence the *strength* of the relationship between a piece of music and one's responses to it (e.g., personality and peer group influences). There is a two-way relationship between personal and situational factors given that, in a high-performance context, the music should be *functional* or carefully coordinated with the tasks and specifics of what is required (Kodzhaspirov, Zaitsev, & Kosarev, 1986). The 'Situational Factors' are subdivided into *environmental factors* that concern aspects of the performance location and how it is set up, and *task-related* factors that entail how the performance is undertaken (e.g., the intensity of an athlete's training session or whether a musical recital is self-led or directed by a conductor).

Personal factors such as musical preferences and attentional style (i.e., associator vs. dissociator) interact with the situational factors to determine a performer's responses to music (Hutchinson & Karageorghis, 2013). For example, an associator who is training to music with a beat that matches their work rate is likely to attempt to use the music as a type of metronome to regulate their movements (task-related factor). As well as dealing with such individual-level responses, the conceptual framework also deals with responses to music at a group level (environmental factor).

Joint action entails coordinated and synchronized actions performed by a number of people who share common goals (Sebanz, Bekkering, & Knoblich, 2006); for example the members of an aerobatic display team. Accordingly, if the playing of a popular piece results in a surge of enthusiasm within a group (e.g., sports team, orchestra, workforce) that relates to the phenomenon of joint action (Sebanz et al., 2006), this surge is likely to have a positive effect on an individual who may not necessarily express a liking for that piece, resulting in a behavioral consequence (i.e., more effort expended). Likewise, if a particular piece is highly preferred by the leader of a group, this preference will have a direct influence on her/his subordinates through the nonverbal communication that the leader will exude (Priest & Karageorghis, 2008).

The literature also suggests that rhythm and tempo of music will lead to differential responses depending on the age and personality profile of performers and the nature of the task in which they are engaged (Clark, Taylor, & Baker, 2012; Crust & Clough, 2006; Deutsch & Hetland, 2012). To select music that is age-congruent, consideration needs to be given to the artist, musical idiom/style, and release date (Karageorghis, Priest, Terry, Chatzisarantis, & Lane, 2006). Selections and musical styles that were popular during a performer's formative years are known to have a particularly powerful effect owing to the formation of preferences and associations during this time (Brattico, Bogert, & Jacobsen, 2013).

In considering the influence of personality on music selection (a personal factor), extraverts should generally prefer more stimulative music than introverts (Eysenck, 1967; McCown, Keiser, Mulhearn, & Williamson, 1997). Such music is characterized by a fast tempo, prominent rhythmical features, and exaggerated bass tones; the 2014 international smash hit *Happy* (160 bpm) by US artist Pharrell Williams would be a particularly good example of a stimulative track. The melodic and harmonic qualities of music will lead to differential responses in the listener depending on their cultural background and musical upbringing. Westerners, for example, typically associate major-scale melodies and harmonies with positive feelings and happiness. Interestingly, in many Eastern cultures (e.g., Arabic and Persian), minor-sounding melodies and harmonies induce a similar response due to cultural differences in musical composition (Levitin, 2006).

Numerous studies have shown that the personal factors of gender and age moderate one's response to music in performance domains (e.g., Crust, 2008; Karageorghis, Terry, & Lane, 1999; Karageorghis, Priest, Williams, Hirani, Lannon, & Bates, 2010; McCown et al., 1997; Priest, Karageorghis, & Sharp, 2004). For example, in an exercise context, females rate the importance of rhythmic qualities and the 'danceability' of music more highly than males do (Karageorghis et al., 1999). However, males value the importance of cultural associations to a greater extent than their female counterparts (Karageorghis et al., 1999). Also, males report a preference for exaggerated bass tones and this may have to do with the fact that the male voice is generally about one octave lower than the female voice; this difference in voice pitch may account for how sound frequencies are processed by the brain (Priest & Karageorghis, 2008; McCown et al., 1997). In a study that examined members of a health club, it appeared that the younger members generally rated music as being more important to them than older members, additionally preferring contemporary and up-tempo selections (Priest et al., 2004). This finding is consistent with the known role of music in facilitating the formation of a cultural identity during adolescence (Tarrant, North, & Hargreaves, 2001). All age groups consider the churn or variety of music programs to be an important factor (Priest et al., 2004); over-familiarity with a particular piece of music can lead to a decrement in liking (see Berlyne, 1971).

Personality type (e.g., extrovert vs. introvert), musical upbringing–which goes hand-in-hand with cultural background–and hearing acuity are factors that are likely to moderate a performer's response to changes in musical dynamics. For example, if you are an extrovert, who grew up listening to heavy metal music, *and* are hard of hearing, it is unlikely that you will find episodes of soft music beneficial in preparation for a performance that requires engagement of the entire musculature such as sprinting or dancing.

Consequences of Music Use

Consequences relate to the main responses and outcomes associated with music use in a performance domain. With reference to the literature in this field (see Karageorghis & Priest, 2012a, 2012b for a review), the two strongest and most consistent sets of consequences appear first (psychological and psychophysical) followed by behavioral consequences and finally psychophysiological consequences, which are the least consistent (see Karageorghis, 2015).

Among the psychological consequences, the literature reveals that music can facilitate performers' perceptions of *flow* states (Pain, Harwood, & Anderson, 2011). Flow is considered an elusive state of mind associated with complete absorption in a given task. Athletes and musicians have reported elevated levels of performance as a consequence of being in a state of flow, wherein their execution of the task at hand becomes almost automatic and effortless (Jackson & Kimiecik, 2008; Sinnamon, Moran, & O'Connell, 2012).

Many scientific studies illustrate how the consequences that are displayed in the model can be experienced in unison. In the exercise field for example, the use of well-selected music use can result in more positive affect that is coupled with greater work output (Elliott, Carr, & Savage, 2004; Karageorghis, Mouzourides, Priest, Sasso, Morrish, & Walley, 2009; Karageorghis et al., 2010). Similarly in the medical field, surgeons expend their energy more efficiently and report feeling calmer (anxiety control), when listening to music in operating theaters (Moris & Linos, 2013).

There is a feedback loop from the consequences back to the music factors that influences future selection decisions based on a performer's experienced outcomes. This feedback loop entails reflection on the consequences of music use with reference to the moderators. If an athlete or musician selects pieces of music that lead to positive consequences (see right-hand side of Figure 2), such selections are more likely to be used again and promote the selection of similar pieces (e.g., works by the same artist, in the same idiom or of a similar tempo/rhythmic feel). The converse also holds, wherein the experience of a negative consequence may lead to the deselection of a particular track or others like it.

There is a pronounced link between the intrinsic music factors and responses. For example, the *temporal* or time-related aspects of music such as tempo, rhythm, and meter can have a strong influence on arousal. Contrastingly, aspects such as melody and harmony are particularly important in terms of evoking an emotional response in the listener (Gabrielsson & Lindström, 2010; Juslin, 2013a). One of the central predictions of the conceptual framework is that the influence of music factors on human responses is moderated by a host of personal and situational factors that those working in high-performance domains should consider in how they select music.

Music Applications in High-Performance Domains

In the remainder of this chapter we will use the theoretical model presented earlier (Figure 2) as a prism through which to explore a range of music-related applications. These applications will focus primarily on the pre-performance use of music, which we deem to be most relevant to the readers of this book (e.g., using music for mental preparation prior to a concert performance). There are, nonetheless, brief sections on the in-performance and post-performance uses of music. The musical examples that are provided are *indicative* and certainly not *prescriptive* given that musical preferences and aesthetics are rather idiosyncratic in nature (Liljeström, Juslin, & Västfjäll, 2012).

The Application of Pre-Performance Music

In high-performance domains, music can play a pivotal role through forming part of a pre-performance routine and creating a mindset that the individual associates with superior levels of performance. Research has shown that music can have a *priming* effect, which means that it can activate certain thoughts in your mind and influence your motivation toward a particular task without your awareness (Goerlich, Witteman, Schiller, van Heuven, Aleman, & Martens, 2012; Loizou & Karageorghis, 2015; Loizou, Karageorghis, & Bishop, 2014). Thus, pre-performance music can be used to manipulate a performer's psychological state (see 'Consequences' box in Figure 2) and provide a bridge toward their higher purposes and goals.

Research has shown that individuals are largely unaware of the processes that underlie their perceptions, goal pursuits, and behaviors (Hagger & Chatzisarantis, 2014). Given that such processes play a pivotal role in high-performance domains, taking some control over them through the measured application of pre-performance music can facilitate a distinct pattern of thoughts and feelings that lead to a sense of personal mastery and control. Music is known to activate the emotional and movement-related segments of the brain such as the amygdala, temporal lobe, and cerebellum (Chanda & Levitin, 2013; Koelsch, 2010, 2014), and can therefore complement mental preparation in a broad range of performance domains.

Scientists maintain that music has a particularly strong influence on the brain's unconscious processes (Levitin & Tirovolas, 2009; Scherer & Zentner, 2001). You will recall from the first half of this chapter that well-selected music can enhance how people feel in a general sense (affect), although it bears little influence on their perceived exertion at high exercise intensities (RPE; Boutcher & Trenske, 1990; Hutchinson & Karageorghis,

Antonio's Big Stage Dream

Antonio is an operatic tenor with major ambitions. Prior to his birth, Antonio's family emigrated from a picturesque village near Naples in Italy to north London in the UK. Antonio's father was a skilled pastry chef who scrimped and saved to open a small bakery. While he worked, he would play his favorite Neapolitan songs in the background; the music conjured fond memories of his homeland and the much-loved relatives he had left behind (see 'Extrinsic Musical Factors' in Figure 2). Sadly, the master baker died prematurely in the early 90s and Antonio's eldest brother was obliged to take over the business. Antonio was just a teenager when his dad passed away but the doleful schoolboy had inherited a strong passion for Italian music. He went on to become one of the best classical singers at the music college in central London where he studied for three years. After graduating from the college, Antonio experienced financial hardship. Despite considerable personal anguish, he gave up trying to be a professional singer, because he could earn much more working for his brother in the bakery.

While at music college Antonio dreamed of joining the musical elite; he idolized the legends of his field such as Luciano Pavarroti and Mario Lanza, and was intent on following in their footsteps. Although he was generally comfortable with small and intimate performances, such as the college's lunchtime concerts or singing at the local Italian church, he was often paralyzed by fear when it came to performing in large concert venues. The nerves were so extreme that Antonio's mouth would dry up, his throat would feel constricted, and his broad frame would be debilitated by severe muscular tension. The impact of the nerves on his performances was all too obvious with him forgetting his words, losing his timing, and sounding 'squeaky' rather than fulsome and well-rounded.

Following graduation, nerves had held Antonio back for over a decade and he had almost grown content with working for his brother and singing in Italian restaurants or at occasional amateur operatic productions. He knew that his voice was much stronger than that of the amateurs who surrounded him—in fact he relished being a big fish in a small pond—nonetheless he retained the fear associated with appearing in the concert halls. Antonio knew that there was a level of irrationality underlying his stage fright but it wasn't until a chance meeting on the subway with his old singing teacher that he resolved to do something about this.

The teacher advised Antonio to use a recording of his favorite artist performing whatever piece or pieces he was to perform that night - this was to be done *en route* to the concert venue as a type of 'pre-performance routine'. While listening to the artist, Antonio was to just 'shadow' the voice, without pushing vocally, and imagine himself performing in the artist's body; dominant and totally self-assured on stage. The *skills* required to sing did not change between the small amateur venue and the large professional venue, so why should the mindset change? Hearing his favorite artist and adopting the persona of a Pavarotti or Lanza filled Antonio with unshakeable confidence and he was able to translate that positive mindset into his performances on stage. With this new approach, Antonio's singing career ascended to a higher plane and a string of paid engagements followed. He still needed to work at the bakery but reduced his hours by half, so that in his mid 30s he had sufficient time to rehearse and rekindle his dream to sing on the big stage.

2013; Karageorghis et al. 2009). To a degree, such findings support the notion that the brain processes music at a subcortical or automatic level without conscious effort. It is precisely this apparent lack of a need for conscious processing that can make music an ideal form of priming for people who wish to enhance their pre-performance routine. In essence, we do not need to think very much for music to influence our behaviors or how we feel.

Both stimulative and calming music has the potential to prepare people for superior performance. The precise choice of music will depend on an individual's desired mindset: for those riddled with anxiety and, in particular, somatic or bodily-type anxiety (as experienced by singer Antonio, see p. 161) the use of calming music can prove beneficial. In contrast, for those who find it difficult to get suitably 'psyched-up' for the big occasion and feel under-aroused, a stimulative selection (see e.g., p. 151) can have the desired effect. This is referred to as *arousal regulation* and represents a psychophysiological consequence of music use (see Figure 2). In some instances, mood maintenance is the key and music that is neither stimulative nor sedative might be used to good effect. The use of music has the potential to influence the various aspects of anxiety. For example, calming music that has strong communal and personal associations (see 'Extrinsic Musical Factors' in Figure 2) can conjure the right type of mental imagery and/or thought processes (heroic images, overcoming adversity, motion-related thoughts, etc.). It may also merely distract us from our inner fears.

At the 2004 Athens Olympics, British middle distance runner Kelly Holmes used the soulful ballads of Alicia Keys such as *Fallin'* and *If I Ain't Got You* as an integral part of her pre-performance routine. Although the music was generally slow in tempo and rhythmically uncomplicated, the uplifting lyrics had an inspirational effect when Holmes was striving to focus on the task at hand and forget about the string of illnesses and injuries that had blighted her preparations. The music also helped Holmes to block out the burden of public expectation that had weighed heavily upon her shoulders. She managed the unprecedented feat, for a British woman, of winning gold medals in both the 800-meter and 1500-meter events.

It is hard to generalize from anecdotal examples such as that of Kelly Holmes given that performers experience anxiety in many different ways. Most people who perform in highly motoric events such as running, throwing, dancing, or weightlifting prefer quite stimulative music with which to prepare mentally. In contrast, those who perform in cognitive performance domains such as music and theatre, public speaking, or open-heart surgery prefer more calming music. When you have to process large amounts of information, engage the memory, or apply fine muscular control (e.g., playing the violin), a calm mental state is associated with superior performance and so the music choices of individuals might reflect this. The point we are trying to make here is that it is certainly not a case of 'one size fits all' and individual performers will need to engage in some self-experimentation to find out what works best for them. Such experimentation is best done around events that are of lesser importance - never leave experimentation for career-defining events such as an Olympic Games or a solo performance at a major international concert venue.

Olympian Michael Phelps' "Listening Bubble"

The American swimmer Michael Phelps has won more Olympic medals than any other athlete in history. He clinched six gold medals in the 2004 Athens Games and surpassed that incredible feat with an unprecedented cache of eight golds at the 2008 Beijing Games. He capped his Olympic career and sealed his name in the annals of sporting history with four golds and two silvers in the 2012 London Games. Phelps creates his own pre-competition "listening bubble" and immerses himself in a distinctly 'rapcentric' playlist. The artists Jay-Z, G Unit, and Eminem are among his favorites. On his pre-race playlist for both Beijing and London was the track *I'm Me* by Lil' Wayne, which includes a lyric about being the best and knowing the game so well that he could be officiating it! The tracks that he chooses have a self-affirming and often aggressive quality that serves to reinforce Phelps' desired mindset for the big occasion.

So what are the characteristics of stimulating and calming music for pre-performance application? Amongst the intrinsic musical factors (see Figure 2), one of the key considerations is tempo or meter. Stimulating music will function to elevate resting heart rate; generally, music with a tempo in excess of 120 bpm is considered stimulative (cf. Karageorghis et al., 1999). Calming music will have a tempo in the range 60-80 bpm and music that is slower than this will have a tendency to send you to sleep (i.e., it is too sedative for high-performance domains). This is because the body's pulses such as heart rate and breathing rate *entrain* to the rhythmical qualities of music and in the instance of music slower than 60 bpm, the cardiovascular system is led toward a level of functioning that is lower than at normal resting levels. Calming music also tends to be rhythmically quite simple; think of slow-tempo classical works by Vivaldi,

Handel, and Bach or for more contemporary examples, tracks by new-age artists Enya or Enigma.

Another important characteristic is instrumentation, with stimulative music being characterized by more percussive, often 'harsher' sounds, such as the driving beat of drums and the raucous rhythmic punctuation of electric guitars. Stimulative music also has pronounced bass frequencies or discernible bass lines that might compel the listener to move in time with the music. The instrumentation used in calming music on the other hand, might be comprised of soothing, 'warm' instruments such as strings, clarinet or piano. The key is for the performer to become absorbed in and comforted by the music, so that it eases the anxiety symptoms that they might be experiencing.

The Application of In-Performance Music

In-performance music is perhaps most applicable to elite athletes who use it as a backdrop for training sessions and for some competitions (e.g., high-jump contests in which each athlete is accompanied by their chosen piece of music). In many other high-performance domains, the nature of the performance–playing an instrument, acting, speaking, etc.–would render the use of music wholly inappropriate. Given the application of in-performance music among athletes, and the fact that there is much that has already been written for this group of performers (e.g., Karageorghis, 2015; Karageorghis & Terry, 2011; Laukka & Quick, 2013; Terry & Karageorghis, 2011), this is a brief subsection that highlights some of the key application principles and provides one example from an endurance sport (see box on p. 166).

One of the principal considerations in the application of in-performance music is whether to use it synchronously or asynchronously (see 'Task-related' box in Figure 2). The synchronous application takes a little more preparation as movement rates for different intensities of activity need to be established–we normally accomplish this using video analysis–with pieces of music selected and segued in accord with the desired work rate. An effective way to engender an ergogenic effect is to select music that is one or two beats per minute faster than the expected work rate. This promotes a slightly higher-than-expected work rate without any perceptible increase in exertion (see e.g., Lim et al., 2014), and serves as an effective technique for use with recreational and elite athletes alike

(Karageorghis et al., 2009, 2010; Simpson & Karageorghis, 2006; Terry, Karageorghis, Mecozzi Saha, & D'Auria, 2012).

Both synchronous and asynchronous music will reduce an athlete's rating of perceived exertion at low-to-moderate work intensities and, if well selected in terms of personal and situational factors (see Figure 2), will also enhance their affective state and cognitions. To maximize the psychological and psychophysical benefits of in-performance music, select tracks with bright, major harmonies, engaging melodic figures, strong rhythmic qualities, and positive lyrical affirmations such as found in the tracks *Harder, Better, Faster, Stronger* by Daft Punk, *Jump To The Beat* by Stacy Lattisaw, or *The Power* by Snap!. An important point to note in the application of asynchronous music is that when a coach is giving instructions, it is best not to use music at all, as it can disrupt one of their essential functions – to apply their expertise in providing feedback to the athlete. Asynchronous music is best used during repetitive and mundane activities that require little verbal instruction such as stretching routines, circuit training, and cardio workouts. For technical sessions, music can be an unwanted distraction.

The Application of Post-Performance Music

It is a given that all performers strive to give the best of themselves on stage, at the lectern, or in the hotbed of competition. The extreme levels of mental and physical effort that can be observed in high-performance domains necessitate effective strategies for recovery and recuperation. In the elite sport domain, recent literature has attached almost as much importance to recovery schedules as to the specifics of training regimens (Kovacs & Baker, 2014; Lunn et al., 2012; Nédélec, McCall, Carling, Legall, Berthoin, & Dupont, 2012, 2013). Given its ability to entrain bodily pulses, to distract the mind and reenergize the body, the judicious application of post-performance music has the potential to aid recovery and recuperation.

The environment is a central consideration when a performer wishes to relax and unwind (see 'Environmental' box in Figure 2). Post-performance music can be easily combined with a sauna, a massage, meditation, yoga, or even a long soak in a hot bath. Similar to calming pre-performance music (60-80 bpm), recuperative music has a slow tempo or meter (60-90 bpm) and is characterized by the sound of warm, soothing instruments; jarring or irritating tracks do not work well where the goal is to calm or sedate. The slightly broader tempo range reflects the fact that in many performance domains the body's pulses are sped up and thus the music will facilitate a gradual and progressive return to resting levels. The inclusion of lyrics is not essential and more a matter of personal preference and situational specificity. For recuperative activities such as yoga or meditation, lyrics may not be desirable so that the level of distraction is minimized allowing the performer to find an inner tranquility. For other types of activities such as a sauna or hot bath, the performer might benefit from the distractive qualities of lyrics and even sing/hum along to maximize the reverie of their experience. Table 1 provides some examples of lyrical recuperative tracks along with some brief analysis.

Cheryl the Triathlete

In her chosen discipline of triathlon, Cheryl excels in cycling and running, but is a relatively weak swimmer. This limitation has held her back from major championship success and is a cause of considerable frustration. She makes a concerted effort to improve by swimming at least 10 miles each week in training, a task that is often lonely and arduous with little variation. Cheryl's coach constantly seeks efficiency gains in her unpolished swimming technique. To this end, a key consideration is to relax and not expend excessive energy through unwanted muscular tension. With this in mind, Cheryl uses a waterproof MP3 player in her swimming cap – the *Speedo Aquabeat.* Her musical preferences tend toward soulful ballads and soft, relaxing music (see 'Personal Factors' box in Figure 2). The R&B diva Mariah Carey is her favorite artist. She associates Carey's uplifting lyrics and vocal dexterity over a five-octave span with the physical dexterity that she needs to excel over the triad of events that comprise the triathlon. The warm harmonies that characterize Carey's music create a positive mood while the flowing rhythms and slower tempi help Cheryl to relax and maintain an efficient and regular swimming cadence. The music provides a particular focus on long-duration swims and distracts her from the pain caused by the lactic acid that accumulates in her arms and legs. Sometimes the lyrical content of the music can provide a significant mental boost and Cheryl feels as though she is being beckoned toward superior performance.

In sequencing a post-performance music program, it is advisable to fully capitalize upon the principles of *entrainment*, wherein bodily pulses attune to qualities of the music, by beginning with tracks slightly above resting heart rate and ending with tracks slightly below it. Adopting this approach enhances the relaxation response and heightens the pleasure of the recuperation or recovery session. The optimal duration of a recovery and recuperation session is 30-60 minutes and it is imperative for any potential distractions to be guarded against during this essential 'me time' (agents, managers, coaches, members of the press, phone calls, etc.). This time affords the perfect forum for an activity that is essential for superior performance: reflection. Music can both shape and guide this reflective experience. By way of example, the celebrated British comedian John Bishop whose most recent tour consisted of 26 live performances in a five-week period, describes how the soothing pieces of David Gray's album *White Ladder* help him to relax and unwind. Bishop revealed to the listeners of BBC Radio 4's *Desert Island Discs* "I played this album to death, it's become part of my DNA [...] when I'm on tour, it's the thing that sort of eases me down, let's me go to sleep. So I listen to *David Gray* probably two or three times a week, put my headphones on, sit back and snooze and feel comforted" (Young & Bishop, 2012).

Table 1: Sample Lyrical Tracks for Recovery and Recuperation

Track Title	Artist(s)	Tempo (bpm)	Analysis
Home Again http://youtu.be/kJ4s3G7hgR4	Michael Kiwanuka	88	This track greets the listener with a rather unusual time signature of 12/8 (twelve quavers per bar arranged in four groups of three). The guitar-led rhythmic structure feels immediately comforting while the rich and soulful voice of Kiwanuka is utterly entrancing.
I've Told You Now http://youtu.be/K3BLywhPiF4	Sam Smith	86	With a warm guitar-led accompaniment, a tender falsetto vocal, and a harmonic structure that flows over you like a summer breeze, this track lulls you into an ever-calmer state.
Maybe Tomorrow http://youtu.be/2q9_ZEtuTR8	Stereophonics	81	A simple harmonic structure and gravelly vocal delivery ensure that the track is immediately absorbing. The lyrical content is quite uplifting, alluding to an idyllic and optimistic state of mind.
Oblivion http://youtu.be/VgXOPeobPcl	Bastille	78	Cushioned by a gentle piano accompaniment, lead singer Dan Smith showcases his angelic voice. When going into falsetto, Smith's vocal has a particularly soothing quality.
I See Fire http://youtu.be/mIIXxyHTzfg	Ed Sheeran	76	The *chanteur de notre temps* at his glorious best; this track pacifies the mind and body with a tender vocal performance over the barest of instrumental accompaniments.
Pachabel: Cannon In D http://youtu.be/ThFe7Qer9Jk	Laura Wright	74	This baroque piece gently guides the heart rate toward resting levels. The articulation of the lyrics has a 'stretchy' quality achieved through a rounded, legato delivery. The harmonies are masterfully overlaid and a gentle ballad-like rhythm holds the track together beautifully.

Track Title	Artist(s)	Tempo (bpm)	Analysis
Don't Leave https://youtu.be/_iOojt3a1Yw	Faithless	72	The defining feature of this track is a wonderful interleaving of male and female vocals. The beat is unashamedly laid back and the rich harmonies nudge the listener toward an ever-deepening state of tranquility.
Clown http://youtu.be/0OHX_PA25Ok	Emeli Sandé	68	The prodigious Sandé soothes the soul with her popular ballad. The vocal tone is breathtaking while the piano-string accompaniment serves as the perfect antidote to the stressors associated with high-performance domains.
If You Wait https://youtu.be/YjtR_zmMbGE	London Grammar	63	A magnificent vocal performance by Hannah Reid is the defining feature of this track. The haunting melody and instrumental simplicity are the keys to its success in facilitating recuperation. The piece might be described as a 'sonic hammock.'
Breathe Me https://youtu.be/Q0QBzX7zZBw	Sia	60	As this ambient track progresses, the listener is eased toward a state of serenity and transcendence. Sia's breathy and somnolent vocals serve to gently caress while transporting you to a land of complete tranquility.

Summary and Recommendations

Music-related interventions can be applied in a multitude of ways in high-performance domains (e.g., to promote task-relevant imagery, enhance self-confidence, and curb anxiety). We have presented a conceptual framework (Figure 2) that illustrates the interaction between the musical stimulus, the nature of the individual/s, and a host of situational factors in determining a performer's or group response to music. Importantly, the decision to select the same type of music for future performances is governed by a feedback loop. We encourage performers to experiment with different musical selections at times when the outcome of the performance is not significant, until they ascertain the pieces that work most effectively for them. These pieces can then be applied in preparation for those performances that might be career-defining. This process is analogous to trying on different items of sports apparel or stage outfits to gauge which makes you feel most comfortable and self-assured.

Music can be used pre-performance as a tool with which to manipulate psychological states. Prior to selecting music, individuals are encouraged to consider their desired mindset for successful performance. Soft, slow music can be used as a sedative and facilitates a calm mental state in performers; such a state is arguably most beneficial for those engaged in tasks that require large amounts of information processing, such as open-heart surgery or the recitation of poetry. Conversely, loud, fast music can be used as a stimulant to increase a performer's level of arousal; such a state is helpful to those who perform highly motoric tasks such as weightlifting or dancing. In addition, music can be used in-performance, either synchronously or asynchronously, although this application is perhaps most relevant to athletes. Used synchronously, music can function like a metronome, as performers can make a conscious attempt to match their movement patterns to the beat, thus enhancing energy efficiency and work output. Used asynchronously, music can lower a performer's perception of exertion and heighten affective states during a given task.

Music has the capacity to entrain movement patterns and the body's pulses without any conscious effort; much in the same way that when pendulum clocks are placed on adjacent walls, the pendulums will, eventually, swing in perfect synchrony with each other. This principle can be applied to many of the functions that music serves. For example, performers are encouraged to consider the post-performance applications, in order to facilitate recuperation from physically and/or mentally demanding performances. A post-performance music program should commence at ~90 bpm and gradually decrease to ~60 bpm (see e.g., Table 1). Just as gravity pulls at a hot air balloon until it returns safely to the ground, so too can the music's calming pulse entrain a performer's heart rate back toward resting levels.

We hope that by applying the principles outlined in this chapter, you will develop a much fuller understanding of how music can be used in high-performance domains. Performers resemble finely-balanced weighing scales; a seemingly delicate push in the wrong direction can cause the scales to come crashing down. The crash of the scales is perhaps an apt metaphor for the phenomenon of 'choking' that is expounded throughout this volume. We encourage performers to use music in a structured and systematic fashion, in order to help tip the scales slightly in their favor. Music has the capacity to calm or excite an individual performer, unite a seemingly disparate group, stir a wave of enthusiasm in a team under duress, and elevate performance levels. Collectively, such benefits can counter the inevitable consequences evoked in this chapter's titular track: *Under Pressure*.

References

Berlyne, D. E. (1971). *Aesthetics and psychobiology.* New York, NY: Appleton Century Crofts.

Boutcher, S. H., & Trenske, M. (1990). The effects of sensory deprivation and music on perceived exertion and affect during exercise. *Journal of Sport & Exercise Psychology, 12,* 167-176.

Brattico, E., Bogert, B., & Jacobsen, T. (2013). Toward a neural chronometry for the aesthetic experience of music. *Frontiers in Psychology, 4,* 1-21.

Chanda, M. L., & Levitin, D. J. (2013). The neurochemistry of music: Evidence for health outcomes. *Trends in Cognitive Sciences, 17,* 179-193.

Clark, I. N., Taylor, N. F., & Baker, F. A. (2012). Music interventions and physical activity in older adults: A systematic literature review and meta-analysis. *Journal of Rehabilitation Medicine, 44,* 710-719.

Clark, T., Lisboa, T., & Williamon, A. (2014). An investigation into musicians' thoughts and perceptions during performance. *Research Studies in Music Education, 36,* 19-37.

Crust, L. (2008). Perceived importance of components of asynchronous music during circuit training. *Journal of Sports Sciences, 26,* 1547-1555.

Crust, L., & Clough, P. J. (2006). The influence of rhythm and personality in the endurance response to motivational asynchronous music. *Journal of Sports Sciences, 24,* 187-195.

Deutsch, J., & Hetland, K. (2012). The impact of music on pacer test performance, enjoyment and workload. *Asian Journal of Physical Education & Recreation, 18,* 6-14.

Elliott, D., Carr, S., & Savage, D. (2004). Effects of motivational music on work output and affective responses during sub-maximal cycling of a standardized perceived intensity. *Journal of Sport Behavior, 27,* 134-147.

Eysenck, H. J. (1967). *The biological basis of personality.* Springfield, IL: Thomas Publishing.

Fritz, T. H., Jentschke, S., Gosselin, N., Sammler, D., Peretz, I., Turner, R., . . . Koelsch, S. (2009). Universal recognition of three basic emotions in music. *Current Biology, 19,* 573-576.

Gabrielsson, A., & Lindström, E. (2010). The role of structure in musical expression of emotions. In P. Juslin & J. Sloboda (Eds.), *Handbook of music and emotion: Theory, research, and applications* (pp. 367-400). New York, NY: Oxford University Press.

Goerlich, K. S., Witteman, J., Schiller, N. O., van Heuven, V. J., Aleman, A., & Martens, S. (2012). The nature of affective priming in music and speech. *Journal of Cognitive Neuroscience, 24,* 1725-1741.

Hagger, M. S., & Chatzisarantis, N. L. D. (2014). An integrated behavior change model for physical activity. *Exercise and Sport Sciences Reviews, 42,* 62-69.

Hutchinson, J. C., & Karageorghis, C. I. (2013). Moderating influence of dominant attentional style and exercise intensity on responses to asynchronous music. *Journal of Sport & Exercise Psychology, 35,* 625-643.

Hutchinson, J. C., Karageorghis, C. I., & Jones, L. (2015). See hear: Psychological effects of music and music-video during treadmill running. *Annals of Behavioral Medicine, 49,* 199-211.

Jackson, S. A., & Kimiecik, J. C. (2008). The flow perspective of optimal experience in sport and physical activity. In T. S. Horn (Ed.), *Advances in sport and exercise psychology* (3rd ed., pp. 377-399). Champaign, IL: Human Kinetics.

Juslin, P. N. (2013a). From everyday emotions to aesthetic emotions: Towards a unified theory of musical emotions. *Physics of Life Reviews, 10,* 235-266.

Juslin, P. N. (2013b). What does music express? Basic emotions and beyond. *Frontiers in Psychology, 4,*596.

Karageorghis, C. I. (2015). The scientific application of music in exercise and sport: Towards a new theoretical model. In A. M. Lane (Ed.), *Sport and exercise psychology* (2nd ed., pp. 274-320). London, UK: Routledge.

Karageorghis, C. I., Hutchinson, J. C., Jones, L., Farmer, H. L., Ayhan, M. S., Wilson, R. C., . . . Bailey, S. G. (2013). Psychological, psychophysical, and ergogenic effects of music in swimming. *Psychology of Sport and Exercise, 14,* 560-568.

Karageorghis, C. I., Mouzourides, D. A., Priest, D. L., Sasso, T. A., Morrish, D. J., & Walley, C. L. (2009). Psychophysical and ergogenic effects of synchronous music during treadmill walking. *Journal of Sport & Exercise Psychology, 31,* 18-36.

Karageorghis, C. I., & Priest, D. L. (2012a). Music in the exercise domain: A review and synthesis (Part I). *International Review of Sport and Exercise Psychology, 5,* 44-66.

Karageorghis, C. I., & Priest, D. L. (2012b). Music in the exercise domain: A review and synthesis (Part II). *International Review of Sport and Exercise Psychology, 5,* 67-84.

Karageorghis, C. I., Priest, D. L., Terry, P. C., Chatzisarantis, N. L. D., & Lane, A. M. (2006). Redesign and initial validation of an instrument to assess the motivational qualities of music in exercise: The Brunel Music Rating Inventory-2. *Journal of Sports Sciences, 24,* 899-909.

Karageorghis, C. I., Priest, D. L., Williams, L. S., Hirani, R. M., Lannon, K. M., & Bates, B. J. (2010). Ergogenic and psychological effects of synchronous music during circuit-type exercise. *Psychology of Sport and Exercise, 11,* 551-559.

Karageorghis, C. I., & Terry, P. C. (2011). *Inside sport psychology.* Champaign, IL: Human Kinetics.

Karageorghis, C. I., Terry, P. C., & Lane, A. M. (1999). Development and initial validation of an instrument to assess the motivational qualities of music in exercise and sport: The Brunel Music Rating Inventory. *Journal of Sports Sciences, 17,* 713-724.

Kodzhaspirov, Y. G., Zaitsev, Y. M., & Kosarev, S. M. (1986). The application of functional music in the training sessions of weightlifters. *Soviet Sports Review, 23,* 39-42.

Koelsch, S. (2010). Towards a neural basis of music-evoked emotions. *Trends in Cognitive Sciences, 14,* 131-137.

Koelsch, S. (2014). Brain correlates of music-evoked emotions. *Nature Reviews Neuroscience, 15,* 170-180.

Kovacs, M. S., & Baker, L. B. (2014). Recovery interventions and strategies for improved tennis performance. *British Journal of Sports Medicine, 48,* 1-5.

Laukka, P., & Quick, L. (2013). Emotional and motivational uses of music in sports and exercise: A questionnaire study among athletes. *Psychology of Music,* 41, 198-215.

Levitin, D. J. (2006). *This is your brain on music: The science of a human obsession.* New York, NY: Dutton Penguin.

Levitin, D. J., & Tirovolas, A. K. (2009). Current advances in the cognitive neurosciences of music. *Annals of the New York Academy of Sciences: The Year in Cognitive Neuroscience 2009, 1156,* 211-231.

Liljeström, S., Juslin, P. N., & Västfjäll, D. (2012). Experimental evidence of the roles of music choice, social context, and listener personality in emotional reactions to music. *Psychology of Music, 41,* 579-599.

Lim, H. B. T., Karageorghis, C. I., Romer, L. M., & Bishop, D. T. (2014). Psychophysiological effects of synchronous versus asynchronous music during cycling. *Medicine & Science in Sports & Exercise, 46,* 407-413.

Loizou, G., & Karageorghis, C. I. (2015). Effects of psychological priming, video, and music on anaerobic exercise performance. *Scandinavian Journal of Medicine and Science in Sports, 25,* 909-920.

Loizou, G., Karageorghis, C. I., & Bishop, D. T. (2014). Interactive effects of video, priming, and music on emotions and the needs underlying intrinsic motivation. *Psychology of Sport and Exercise, 15,* 611-619.

Lunn, W. R., Pasiakos, S. M., Colletto, M. R., Karfonta, K. E., Carbone, J. W., Anderson, J. M., & Rodriguez, N. R. (2012). Chocolate milk and endurance exercise recovery: Protein balance, glycogen, and performance. *Medicine & Science in Sports & Exercise, 44,* 682-691.

McCown, W., Keiser, R., Mulhearn, S., & Williamson, D. (1997). The role of personality and gender in preference for exaggerated bass in music. *Personality and Individual Differences, 23,* 543-547.

Moris, D. N., & Linos, D. (2013). Music meets surgery: Two sides to the art of "healing". *Surgical Endoscopy, 27,* 719-723.

Nédélec, M., McCall, A., Carling, C., Legall, F., Berthoin, S., & Dupont, G. (2012). Recovery in soccer: Part I – Post-match fatigue and time course of recovery. *Sports Medicine, 42,* 997-1015.

Nédélec, M., McCall, A., Carling, C., Legall, F., Berthoin, S., & Dupont, G. (2013). Recovery in soccer: Part II – Recovery strategies. *Sports Medicine, 43,* 9-22.

Neil, R., Wilson, K., Mellalieu, S. D., Hanton, S., & Taylor, J. (2012). Competitive anxiety intensity and interpretation: A two-study investigation into their relationship with performance. *International Journal of Sport & Exercise Psychology, 10,* 96-111.

North, A. C., & Hargreaves, D. J. (2008). *The social and applied psychology of music.* Oxford, UK: Oxford University Press.

Pain, M. A., Harwood, C., & Anderson, R. (2011). Pre-competition imagery and music: The impact on flow and performance in competitive soccer. *The Sport Psychologist, 25,* 212-232.

Priest, D. L., & Karageorghis, C. I. (2008). A qualitative investigation into the characteristics and effects of music accompanying exercise. *European Physical Education Review, 14,* 347-366.

Priest, D. L., Karageorghis, C. I., & Sharp, N. C. C. (2004). The characteristics and effects of motivational music in exercise settings: The possible influence of gender, age, frequency of attendance, and time of attendance. *Journal of Sports Medicine and Physical Fitness, 44,* 77-86.

Sanchez, X., Moss, S. L., Twist, C., & Karageorghis, C. I. (2014). On the role of lyrics in the music–exercise performance relationship. *Psychology of Sport and Exercise, 15,* 132-138.

Scherer, K. R., & Zentner, M. R. (2001). Emotional effects of music: Production rules. In P. N. Juslin & J. A. Sloboda (Eds.), *Music and emotion: Theory and research* (pp. 361-392). Oxford, UK: Oxford University Press.

Sebanz, N., Bekkering, H., & Knoblich, G. (2006). Joint action: Bodies and minds moving together. *Trends in Cognitive Sciences, 10,* 70-76.

Simpson, S. D., & Karageorghis, C. I. (2006). The effects of synchronous music on 400-m sprint performance. *Journal of Sports Sciences, 24,* 1095-1102.

Sinnamon, S., Moran, A., & O'Connell, M. (2012). Flow among musicians: Measuring peak experiences of student performers. *Journal of Research in Music Education, 60,* 6-25.

Stork, M. J., Kwan, M., Gibala, M. J., & Martin Ginis, K. A. (2014). Music enhances performance and perceived enjoyment of sprint level interval exercise. *Medicine and Science in Sport and Exercise, 47,* 1052-1060.

Tarrant, M., North, A. C., & Hargreaves, D. J. (2001). Social categorization, self-esteem, and the estimated musical preferences of male adolescents. *The Journal of Social Psychology, 141,* 565-581.

Terry, P. C., & Karageorghis, C. I. (2011). Music in sport and exercise. In T. Morris & P. C. Terry (Eds.), *The new sport and exercise psychology companion* (pp. 359-380). Morgantown, WV: Fitness Information Technology.

Terry, P. C., Karageorghis, C. I., Mecozzi Saha, A., & D'Auria, S. (2012). Effects of synchronous music on treadmill running among elite triathletes. *Journal of Science and Medicine in Sport, 15,* 52-57.

Young, K. (Presenter), & Bishop, J. (Guest). (2012, June 24). In L. Buckle (Producer), *Desert island discs.* London, UK: British Broadcasting Corporation, Radio 4.

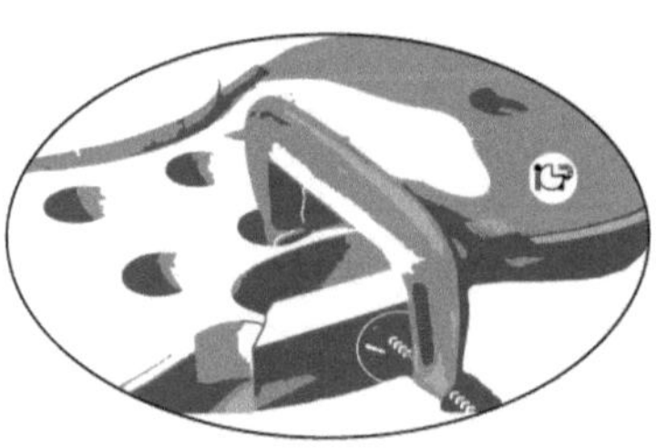

The Impact of Listening to Music on Musicians' Practice and Performance[1]

Music inspires and empowers people. By making music we are able to get in touch with our emotions, desires, sorrows, our joy and our dreams – everything that is important to us as sentient beings. By listening to music we may close our eyes and forget about the outer world. Thus, music enables the opportunity to feel comfortable, either by stimulating us or by calming us down. As presented in the preceding chapter "Under Pressure: Music-Related Interventions in High-Performance Domains"(-Karageorghis & Bird), the power of music plays a vital role in high-performance domains like sports, business, and music itself. The authors introduce a heuristic theoretical framework to demonstrate the impact of music applications particularly in pre-performance settings. According to them, carefully chosen and arranged music applications can foster the performer in achieving an optimal or peak performance by influencing one´s mindset prior to a performance or activity.

It is a long and hard journey to become a professional musician who can stand one's ground in the world of high-performance music. One usually starts playing an instrument in (early) childhood. It requires more than 10,000 hours of deliberate practice to reach a level of expertise (Ericsson, 2006). And once one has passed a performance and qualification test at the music conservatoire, daily challenges are to be faced and mastered. After graduation, the musician's professional life is affected by daily practice, teaching, performing, and the "hunt" for new engagements. Strong discipline, a high level of frustration tolerance, and physical and mental health are essential elements of life as a professional musician who achieves performance excellence – often under pressure. Even though there is evidence that music promotes physical and mental health (Rickard & McFerran, 2012), recent research reveals that continual stress and music performance anxiety (MPA) cause a significant threat to the wellbeing of musicians (Ginsborg, Spahn & Williamon, 2012; Kenny, 2011; Kreutz, Ginsborg & Williamon 2008; Williamon & Thomson, 2006). According to Osborne et al., conservatoire students, in particular, seem to suffer to a great extent from MPA. Their recent investigation

> [...] demonstrates that musicians who learn performance psychology skills in a conservatoire setting are not only able to better regulate and manage perfor-

1 Comment on Karageorghis and Bird's chapter in this volume, p. 149ff.

> mance anxiety, they also improve in a range of skills associated with psychological resilience and wellbeing in music performance (2014, p.14).

How, furthermore, a conservatoire student's motivation and practice behavior can have an impact on wellbeing is another interesting question of a recent study. To provide efficient self-regulated strategies for a successful, meaningful, and healthy life as a professional high-performance musician represents the principal aim of a recent investigation (Spangler, manuscript in preparation).

Motivation to perform can be positively influenced through the impact music has on mindset (see Karageorghis & Bird in this volume), even if it is just for a short time. It is commonly accepted that optimal physical and psychological wellbeing, as well as physical and mental health, are essential in order to utilize one's capabilities in expert performance (Ackermann et al., 2012; Kenny 2011; Williamon 2004). According to Ryan and Deci (2000b), the satisfaction of psychological needs leads to optimal wellbeing. Hence, in order to accomplish and maintain wellbeing on a long-term basis, it is necessary to satisfy the psychological need for competence, autonomy, and social relatedness. Referring to competence and autonomy, self-regulated practice plays a crucial role for expert musicians. Strategies like playing a piece in different ways to challenge oneself, e.g., using diverse coordinative, cognitive, or (senso)motorical skills, help the performer to deepen and internalize the technical and musical skills needed to achieve an artistically high performance level (Greene 2012a). In addition, substantial key qualifications such as self-esteem, self-confidence, and self-efficacy are promoted by developing these competences.

From a long-term view, one might question whether a professional musician who is prepared optimally and has already reached a high level of expertise, really needs a pre-performance routine like listening to music. And if somebody is not prepared optimally, can listening to a particular piece of music in pre-performance increase the level of the performance that follows at all? While MacDonald et al. (2003) discovered that listening to music as a strategy to cope with physical and mental stress is generally effective, it always depends on the level of perceived stress. In situations of heavy and intense stress, the musical intervention was not effective. Hence, in terms of listening to music, further investigations should be conducted that also examine the level of MPA in pre-performance.

Pre-Performance Routines as a Strategy to Enhance Performance

Rock climbing requires skills as diverse as endurance, strength, and coordination, but also mental capabilities. Even optimally prepared, high-level climbers may be anxious about failure or injury. Thus, they employ self-regulation strategies, as demonstrated in this example:

In an interview, a professional rock climber described how her ritual prior to stepping into a requested route evolved over time. She always used the same routine in the same order: put on the rope, sit down, put on the left climbing shoe then the right one, get up, take off the sport jacket, stretch both arms (see power pose), and take a deep breath. It was only after all these procedures that she would step onto the wall. She got used to the routine, and it worked excellently and enabled her to stay mentally calm and focused, but also physically activated. One day the ground was still wet from rain and there was no rock or tree trunk around to sit on, so she couldn't sit down to put on her climbing shoes. As a result she got nervous and anxious, lost self-confidence, and eventually failed. The task wasn't too difficult for her, but she relied on the rituals instead of on her own sense of competence and extensive experience.

The widely accepted principle of psychological conditioning explains this phenomenon, as it is transferable into every domain such as music and business. The climber had conditioned herself to perform well only following the ritual. Similarly, pre-performance routines may in principle underpin or even manipulate the mindset in a supportive manner – but can it be dangerous to count on them as a key to success? Or is it better to make a unique decision regarding one's pre-performance strategy, in every single occasion, instead of relying on the expected feeling of security?

Karageorghis and Bird point to the pre-performance use of music in high-performance domains such as music and sports, referring to possible priming effects resulting from a pre-performance routine. Numerous research projects have provided evidence that listening to a piece of music can lead to a short-term improvement in cognitive and creative performance. However, this effect can also be provoked by any other, non-musical stimulus that is perceived positively (Degé & Schwarzer 2015). Amy Cuddy's successful concept of the *power pose*[2], might also be considered as a pre-performance strategy by performers and researchers (Cuddy et al. 2012; see also: Carney et al. 2010). Would listening to music in a pre-performance setting be as or more effective than taking a power pose, or the other way round? Or would both strategies linked together produce better or more significant results? Then there is the question of habituation. Repeated use of listening to the same, specific piece of

2 A behavioral psychology method of priming the psyche of the performer for success through an assertive stance ("fake it 'til you make it").

music prior to a performance might reduce the emotional reaction to this stimulus, and thus, also the desired manipulation of the mindset.

In a recent investigation by Osborne et al. (2014) of the impact of skills training in performance psychology on conservatoire students (see pp. 93-114 of this volume), the authors highlight the most effective clinical intervention programs, which consist of diverse behavioral techniques, including relaxation and mental skills training. According to this, the pre-performance use of music is definitely considered one of many possible MPA interventions. Further studies involving more than a short-term manipulation of the musician's mindset will be necessary in order to determine how professional musicians can be supported to achieve their own level of peak performance on a long-term basis.

In reference to the priming effect, Karageorghis and Bird conclude that "[...] music can be used to manipulate a performer's psychological state [...] and provide a bridge toward their higher purposes and goals" (p. 160). Thus, emotional reactions evoked by listening to music can influence the mindset as well as the body (Altenmüller & Bernatzky, 2015).

A similar phenomenon was reported by Grewe et al. (2009; 2009-2010). Their research on "chill" reactions (strong emotional reactions like goose bumps or shivering) of choir singers listening to an emotionally intense choir piece for 30 minutes showed that 72% of the test participants experienced the chill reaction or response. However, after repetition over seven days, the reactions became irregular and infrequent, indicating a habitual effect and revealing that chill reactions are elusive and difficult to replicate. Moreover, the context in which music is perceived plays a crucial role for the quality and quantity of chill reactions. Albeit research has discovered inter-individual reaction patterns, it is evident that these patterns are highly dependent on intra-individual factors (Altenmüller & Bernatzky, 2015). Musical factors such as tempo, meter, acoustic color or orchestration can influence the priming effect (as described by Karageorghis & Bird) just as well as chill reactions (Altenmüller & Bernatzky, 2015). However, Altenmüller and Bernatzky (2015) also refer to further research on this topic that confirms that despite the use of strong emotional music, chill reactions can happen infrequently and are difficult to replicate. Even different combinations of musical factors do not necessarily induce intense reactions. These findings provide incentive for additional empirical research with respect to the significance of the choice of instruments in the context of the pre-performance use of music. Karageorghis and Bird also stress the importance of instrumentation to attain a comfortable physical condition. A study designed to verify the perceptual categories of instruments (e.g., piano is a "warm" instrument, Karageorghis & Bird, p. 164) would be valuable, specifically in relationship to individual bias and needs with respect to professional musicians. Interestingly, there is evidence that anticipation and expectation have a strong impact on the elicitation of emotions. For instance, if certain expectations are fulfilled, emotional satisfaction is likely – even though habituation might occur. And if not, negative emotions can be the consequence, but also laughter, amazement, or goose bumps (Altenmüller & Bernatzky, 2015). This again shows the importance of the variability of the context in which the state of mind is influenced while listening to music. It would be of interest if further studies will be able to provide insight into similarities and analogies between the priming effect and chill-reactions.

Perception of music is one final point worth discussion here. Attention should be paid to the fact that musicians and non-musicians perceive music differently. Jäncke (2015) shows that the neuronal processing and responding of a musician's auditorial cortex while listening to music is higher and stronger than that of non-musicians. Furthermore, he points out that the neuronal reaction of musicians is also very specific. The cortical response of string players, for example, listening to the sound of string instruments is much stronger than the response to listening to brass instruments (Pantev, Roberts, Schulz, Engelien & Ross, 2001). Altenmüller (2003) demonstrates that lay listeners experience music holistically while professional musicians and trained listeners are additionally capable of differentiating specific features and musical factors like the structure of a music piece, stylistic features, and orchestration. Regarding the use of music as a pre-performance strategy for professional musicians, it may be deemed necessary that these differences should be taken into account in further research.

In conclusion, Karageorghis & Bird's chapter "Under Pressure" provides impetus for pre-performance music applications in music, sports, and business. The authors' heuristic and theoretical framework suggests that the carefully chosen application of music can aid the performer to achieve an optimal performance, by influencing her/his mindset prior to the activity. In order to perform excellently *under pressure* in the high-performance domain of music, multi-faceted practice behavior is indispensable. It remains an unanswered question, whether such music applications can be efficacious for performing musicians themselves, especially in cases involving extremely high stress levels. Additionally, it would be important to investigate the way in which these interventions may have a lasting impact on MPA. Above all, the transfer of these research results to musicians should take into account that professional musicians perceive music differently from lay listeners. Taking this into consideration, the concept of using music as a pre-performance strategy in the high-performance domain of sports, may not be easily transferable to the domain of music. Yet, because of the prospect of this potential "performing under pressure" coping strategy, "Under Pressure" provides music psychologists and researchers in related fields with some very interesting questions for further investigation.

References

Ackermann, B., Driscoll, T., & Kenny, D. T. (2012). Musculoskeletal pain and injury in professional orchestral musicians in Australia. *Medical Problems of Performing Artists, 27(4)*, 181–187.

Altenmüller, E. (2001). How many music centers are in the brain? *Annals of the New York Academy of Sciences, 930*, 273-280.

Altenmüller, E., & Bernatzky, G. (2015). Musik als Auslöser starker Emotionen. In Bernatzky, G. & Kreutz, G. (Eds.). *Musik und Medizin. Chancen für Therapie, Prävention und Bildung.* Wien: Springer-Verlag.

Bangert, M. & Altenmüller, E. (2003). Funktionelle und disfunktionelle Plastizität bei Musikern. *Musikphysiologie und Musikermedizin, 10(3)*, 131-148.

Carney, D. R., Cuddy, A. J. C., & Yap, A. J. (2010). Power posing: Brief nonverbal displays affect neuroendocrine levels and risk tolerance. *Psychological Science 21(10)*, 1363–1368

Cuddy, A. J. C, Wilmuth, C. A., & Carney, D. R. (2012). *The benefit of power posing before a high-stakes social evaluation.* Working Paper. Berkeley: University of California.

Degé, F., & Schwarzer, G. (2015). Musik und kognitive Entwicklung. In Kreutz, G. (Ed.), *Musik und Medizin.* 359-374.

Ericsson, K. A. (2006). The influence of experience and deliberate practice on the development of superior expert performance. In Ericsson, K. A., Charness, N., Feltovich, P. J., & Hoffmann, R. R. (Eds.), *The Cambridge handbook of expertise and performance.* Cambridge: Cambridge University Press.

Ginsborg, J., Spahn, C., & Williamon, A. (2012). Health promotion in higher music education. In R. MacDonald, G. Kreutz & L. Mitchell (Eds.), *Music, health & wellbeing.* Oxford: Oxford University Press.

Greene, D. J. (2012a). 11 strategies for audition and performance success: A workbook for musicians. http://psi.dongreene.com/. Accessed 08 January 2016.

Grewe, O. Altenmüller, E., Nagel, F., & Kopiez, R. (2009-2010). Individual emotional reactions towards music: Evolutionary-based universals? *Musicae Scientiae (Special Issue)*, 261-287.

Grewe, O., Kopiez, R., & Altenmüller, E. (2009). Listening to music as a re-creative process: Physiological, psychological and psychoacoustical correlates of chills and strong emotions. *Music Perception, 24(3)*, 297-314.

Jäncke, L. (2015). Musik und Hirnplastizität. In Bernatzky, G. & Kreutz, G. (Eds.). *Musik und Medizin. Chancen für Therapie, Prävention und Bildung.* Wien: Springer-Verlag.

Kenny, D. T. (2011). *The psychology of music performance anxiety.* Oxford: Oxford University Press.

Kreutz, G., Ginsborg, J., & Williamon, A. (2008), Music students' health problems and health-promoting behaviours. *Medical Problems of Performing Artists, 23*, 3–11.

McDonald, R. A. R., Mitchell, L. A., Dillon, T., Serpell, M. G., Davies, J. B., & Ashley, E. A. (2003). An empirical investigation of the anxiolytic and pain reducing effects of music. *Psychology of Music, 31*, 187–203.

Osborne, M. S., Greene, D. J., & Immel, D. T. (2014). Managing performance anxiety and improving mental skills in conservatoire students through performance psychology training: a pilot study. *Psychology of Well-Being: Theory, Research and Practice, 4(18)*, 2-17.

Pantev, C., Roberts, L. E., Schulz, M., Engelien, A., & Ross, B. (2001). Timbre-specific enhancement of auditory cortical representations in musicians. *Neuroreport, 12*, 169–174.

Rickard, N. S., & McFerran, K. (Eds.). (2012). *Lifelong engagement with music: Benefits for mental health and well-being.* New York: Nova.

Ryan, R. M. & Deci, E. L. (2000b). Self-determination theory and the facilitation of intrinsic motivation, social development, and wellbeing. *American Psychologist, 55*, 68-78.

Spangler, S. (in preparation). *How to stay confident: Motivation, wellbeing and practice behavior among conservatoire music students.* Dissertation. Munich: University of Music and Performing Arts.

Williamon, A. (Ed.). (2004). *Musical excellence: Strategies and techniques to enhance performance.* Oxford: Oxford University Press.

Williamon, A., & Thompson, S. (2006). Awareness and incidence of health problems among conservatoire students. *Psychology of Music, 34(4)*, 411–430.

Afterword: *da capo al fine*

The task of an editor is to choose those contributions that best illuminate a subject and arrange them so that the presentation of thought and information flows. The aim of this book is to display a wide-angle view of the topic "Performing Under Pressure." Therefore, the preceding pages were authored by a heterogeneous group of authors, whose research methods, results, and theories are meant to inspire the reader to think critically, and perhaps motivate him/her to try to seek out and apply new strategies or conduct further research. If this mission is accomplished, there should be nothing left to be said at the end of this book. Or is there?

As a musician, I'm truly fond of the three-part form. While they are not marked as such, this book follows the structure and logic of A – B – A', in which the last section repeats themes of the first in a slightly different way. The "A" part of this book, although meant as an introduction, is so rich in material that it would be a shame to forgo this chance to return to it, especially in the light of the chapters that followed. Perhaps it is the part of my DNA that motivates the teacher in me, combined with my sense for musical composition, that encourages me to return to the beginning to revisit and emphasize the voyage the book embarked on, as well as to find a fitting end (i.e., *fine*) to this Volume III of the *Art in Motion* series.

In his Foreword, Bernd Redmann addresses several of the most prominent features of stage fright: nervousness in advance of the event, concerns about the audience's evaluation of his performance, and a drive to perfectionism leading to worry that he will fall short or fail, along with physical and behavioral symptoms that accompany his cognitive and emotional signs of stress. Redmann describes the task of public speaking in a foreign language; yet studies show that presenting oneself in front of an audience – even in one's native tongue – can cause more anxiety than the fear of death (Marshall, 1994). It's simply human nature to feel stress when "performing under pressure."

One questions why millennia of human existence and experience as a social animal haven't led to better coping strategies. At the *Art in Motion* 2014 symposium, John Wallace named his keynote speech "Confessions of a Justified Sinner." Wallace, Principal of the Royal Conservatoire of Scotland and famous for his trumpet solo at Charles, Prince of Wales, and Lady Diana Spencer's wedding (live broadcast with an estimated 750 million viewers), claimed that he is *not* a natural born performer, but "a quivering mass of self-deprecation" who wonders whether "self-torture" is necessary, or whether there is another way to perform (Wallace, 2014).

Not every performer is willing to discuss his or her self-talk – self-assuring or self-deprecating – as openly as Wallace did. This is just one of many aspects of Ingolf Turban's and Richard A. Schmidt's personal reflections that makes both of these essays rare and precious. Here are two men whose own stories emphatically and accurately identify themes common to the entire volume and to all forms of the arts and sport:

- Pressure is a universal experience in almost all performance situations.
- Preparation *before* is as essential as thoughts *during* performance.
- One can only perform what one has rehearsed.
- Control is more an illusion than a virtue.

It's the responsibility of teachers to address stage fright with their students, *before* stage fright becomes part of their experience.

Given the prevalence of performance anxiety across domains and despite widespread interest and current research methodology, there is still no one unified school of thought as to the cause and effect when it comes to human behavior that ranges from moments of high-achievement to experiences of choking under pressure. In the chapters that follow Redmann, Turban, and Schmidt, that is, in the main, or "B" section of this book, an exploration of causes, effects, and new strategies are presented. One of the goals of both the symposia and the book series is to motivate collaboration and, ultimately, lead to a new understanding of artistic and athletic motor learning and performance. This *raison d'être* for *Art in Motion* is demonstrated by the frequent "to do" lists at the end of the chapters in the form of recommendations for further research and experimentation.

"You can change your life in a dance class!" says Royston Maldoom, choreographer and dance teacher, in the documentary film *Rhythm Is It*. Indeed, one question posed before embarking on such an endeavor as *Performing Under Pressure*, is "Can you change your life by reading a book?" I believe that I am summing up the intentions of all of the individuals involved in this book, when I write that reading alone will not be enough to make a difference, but equipped with knowledge and a variety of perspectives provided here, one can modify one's own practice, pre-performance, and performance rituals and thus, yes, effect change.

Postscript

He was never short on creativity and constructive criticism, nor shy on patience. Without the input and influence, as well as the subliminal motivation and additional help from Detlef Levin, this book would never have reached its excellence as a finished product that is a pleasure for both the eye and the mind.

Additional heartfelt thanks go also to Tim Lee and Gwen Gordon for their help finding an appropriate way to honor and remember Richard A. Schmidt, whose quick-witted humor, curiosity, and intelligence will remain inspirational through his writings. We will miss him dearly. Fortunately, he completed and edited his contribution, which is a crown jewel in this book. "From Heartbeat to Postural Control: Stability is not a State but a Process" would have been a chapter here, had not the brilliant and innovative Josef Ilmberger, long-term champion of the *Art in Motion* idea and invaluable contributor to the symposia, passed away unexpectedly in late 2014. Both Dick and Joe will remain in our hearts as champions of interdisciplinary thought.

References

Grube, T., & Lanch, E. S. (Directors). (2004). *Rhythm is it* [Motion picture on DVD]. Germany: Boomtown Media.

Marshall, J. R. (1994). *Social phobia: From shyness to stage fright.* New Yorks: BasicBooks.

Wallace, J. (2014). Confessions of a justified sinner [Abstract]. *Performing under pressure. Symposium program*, p. 22.

Sebastian Altfeld works in the Unit of Sport Psychology at the Faculty of Sport Science at Ruhr-Universität Bochum (Germany). Sebastian graduated with a master's degree in clinical psychology and has published in the area of coaches' burnout during his PhD studies. Beside his research, he focuses on the work with athletes, coaches, and referees in different kind of sports as a trained sports psychologist. He has been admitted to the pool of sport psychological experts of the German Federation of Sport Science as well as to the European Network of Young Specialists in Sport. Moreover, he is part of the team of sport psychologists in the project www.mentaltalent.de, which offers sport psychology to young talented athletes, their coaches, and parents in Germany.

sebastian.altfeld@rub.de

Lisa Aufegger is a PhD candidate, supported by a Stanbridge Drake-Brockman Award, at the Royal College of Music, London, and worked as a Peter Sowerby Research Associate in Performance Science. She graduated in psychology at the University of Klagenfurt, Austria, where she studied perceptual skills in musicians and non-musicians. Her current research focuses on music performance anxiety (MPA) and specifically the applied psychophysiology of MPA. She is also interested in simulated environments and smartphone applications and their use for performance training. Her work in this area was supported by the Arnold Bentley New Initiatives Fund (SEMPRE) and awarded at the Musicians' Health & Performance 2nd Conference, MHPC 2015, in Odense, Denmark.

lisa.aufegger@rcm.ac.uk

Jonathan Bird is a PhD student in sport and exercise psychology in the Department of Life Sciences at Brunel University London. He is conducting research into the efficacy and application of audiovisual stimuli in the exercise context. He has presented his work at scientific conferences and currently has a number of journal articles under review. Jonathan has a strong interest in applied sport psychology and is accredited as a Sport and Exercise Scientist with the British Association of Sport and Exercise Sciences. He is also a chartered member of the UK's Science Council and has worked as a consultant with athletes from a wide variety of sports. In his spare time, Jonathan enjoys long-distance running and has competed in events that span 5 km to the marathon distance.

jonathan.bird@brunel.ac.uk

K. Anders Ericsson, PhD, is presently Conradi Eminent Scholar and Professor of Psychology at Florida State University. He collaborated with the Nobel Prize winner in Economics, Herbert A. Simon, on verbal reports of thinking leading to their classic book "*Protocol Analysis: Verbal Reports as Data*" (1984). Currently he studies the measurement of expert performance in domains, such as music, chess, and sports, and how expert performers attain their superior performance by acquiring complex cognitive mechanisms through extended deliberate practice. He has edited the influential "*Cambridge Handbook of Expertise and Expert Performance"* and "*Development of Professional Expertise*." He is a Fellow of the *Center for Advanced Study in the Behavioral Sciences*, of the *American Psychological Association*, the *Association for Psychological Science* and a member of *Royal Swedish Academy of Engineering Sciences*.

ericsson@psy.fsu.edu

Paul Evans is a Senior Lecturer in educational psychology at the University of New South Wales in Sydney, Australia. He obtained his undergraduate degree in music education with honours at UNSW, a master's degree in music education at the University of Western Australia, and a PhD in music education at the University of Illinois at Urbana-Champaign. For the latter, he held the Marilyn Pflederer Zimmerman doctoral fellowship and completed his research under the supervision of Gary McPherson. Paul's research interests span a range of topics including motivation and self-regulated learning in educational settings, with an emphasis on the context of instrumental music learning from childhood through to adulthood.

paul.evans@unsw.edu.au

K. Wolfgang Kallus is the Head of Work-, Organisational and Environmental Psychology at the Department of Psychology in the faculty of Natural Sciences of the University of Graz (Austria). His research in the past years focused on stress and regeneration especially in aging workers, applied methods, and Human Factors with a focus on Aviation Psychology. He is a member of various national and international scientific organisations, past president on the board of the IEA-Group "Psychophysiology in Ergonomics (PIE)," president of the Austrian Aviation Psychology Association, and since 1995 director of the Institute for Evaluation Research (Begleitfoschung). In 2012 he received the biannual award of the European Association of Aviation Psychology.

wolfgang.kallus@uni-graz.at

Dr Costas Karageorghis is a Reader in Sport Psychology in the Department of Life Sciences at Brunel University London. He has worked with many UK national governing bodies of sport as a consultant psychologist, as well as with a number of multinational companies that include Nike, Red Bull, Sony, Spotify, Speedo, and the International Management Group (IMG). He is best known for his research and applied work in the area of music-related interventions in exercise and sport. Costas's scientific output includes more than 150 scholarly articles, ten book chapters, and two textbooks including, most recently, *Applying Music in Exercise and Sport* (Human Kinetics, 2016). During his spare time, Costas enjoys playing piano in the jazz duo *Blue Rondo* alongside drummer Joel Shopland.

costas.karageorghis@brunel.ac.uk

Michael Kellmann is Professor of Sport Psychology at the Faculty of Sport Science at Ruhr-Universität Bochum (Germany) and an Honorary Professor at the School of Human Movement and Nutrition Sciences at The University of Queensland (Australia). He is a member of the Association of Applied Sport Psychology, the German Psychological Association, and the Australian Psychological Society. Michael's research activities include overtraining prevention and recovery enhancement, sport psychological diagnostics and intervention, coach behaviour during competition and practice, as well as personality and performance competence of coaches in sports.

michael.kellmann@rub.de

Elisa Kupers is an assistant professor at the University of Groningen, the Netherlands, where she works at the department of Special Needs Education and Youth Care. She completed her bachelor's and master's degree in Psychology at the same university. In 2014, she obtained her PhD in Developmental Psychology on the topic of socially situated learning in one-to-one music lessons, after which she worked as a postdoctoral researcher at the Universität Bremen, Germany (department of Musicology and Music Education). Her research interests include learning through interaction, observational research, creativity and motivation in (music) educational contexts.

w.e.kupers@rug.nl

Andreas C. Lehmann is professor of Systematic Musicology and Music Psychology at the Hochschule für Musik Würzburg (Germany). He has written *Psychology for Musicians* (2007, Oxford University Press; with John Sloboda & Robert Woody), edited *Musikpsychologie: Das neue Handbuch* (2008, Rowohlt; with Reinhard Kopiez & Herbert Bruhn), and authored chapters as well as articles in peer-reviewed journals. His main research interests are expertise/expert performance, skill acquisition, and empirical music education. Currently, Dr. Lehmann is vice-president of the German Music Psychology Society (*DGM*), associate editor of the European journal *Musicae Scientiae*, and editorial board member of the *Journal of Research in Music Education.*

ac.lehmann@hfm-wuerzburg.de

Gary E. McPherson studied music education at the Sydney Conservatorium of Music, before completing a master of music education at Indiana University, a doctorate of philosophy at the University of Sydney and a Licentiate and Fellowship in trumpet performance through Trinity College, London. He is the Ormond Professor and Director of the Melbourne Conservatorium of Music at the University of Melbourne, and previously held a position as the Marilyn Pflederer Zimmerman endowed chair in music education at the University of Illinois at Urbana Champaign. He has served as National President of the Australian Society for Music Education and President of the International Society for Music Education. His most important research examines the acquisition and development of musical competence, and motivation to engage and participate in music from novice to expert levels.

g.mcpherson@unimelb.edu.au

Dr. Christopher Mesagno is a Senior Lecturer in the Faculty of Health at Federation University Australia located one hour west of Melbourne, Australia. Dr. Mesagno was the recipient of the 2007 Association for Applied Sport Psychology Dissertation Award, which introduced a self-presentation model of choking under pressure in sport. Dr. Mesagno has presented numerous internationally recognised papers on stress and anxiety in sport performance, which included the opening keynote presentation for the 2014 German Association for Sport Psychology Conference in Munich, Germany, where he suggested that of collaborative research should be undertaken between sport and music researchers. Dr. Mesagno has also successfully developed theory-matched interventions to improve performance of "choking-susceptible" athletes.

c.mesagno@federation.edu.au

Adina Mornell, classical pianist and recording artist, studied Music, American Literature, Musicology, and Psychology, receiving degrees in the United States, Germany, and Austria, the latter being a doctorate in Music Psychology. In addition to empirical research on musical performance, her interests include the study of deliberate practice, attentional focus, and the intersection of neuroscience and music psychology. In 2008, Adina initiated the biannual series of international and interdisciplinary symposia: *Art in Motion. Musical and Athletic Motor Learning and Performance*. She publishes articles on the arts, science, and society, both in English and in German, and has had several CDs released worldwide. She is professor at the University of Music and Performing Arts in Munich.

adina.mornell@hmtm.de

Dr. Margaret Osborne is a Postdoctoral Research Fellow and Lecturer in the Melbourne Conservatorium of Music and Melbourne School of Psychological Sciences at The University of Melbourne. She is also a registered consulting psychologist, occupational rehabilitation and careers counselor, and past-President of the Australian Society for Performing Arts Healthcare. Margaret's work specializes in performance science and psychology. She works on Australian Research Council funded-research examining motivation and practice quality in elite music performance, as well as research and teaching in psychological strategies to manage anxiety and optimize performance in musicians and athletes. Margaret is passionate about integrating cognitive, behavioral and neuroscientific research to formulate best-practice methods to build health and resilience, and maximize performance potential.

mosborne@unimelb.edu.au

Amelia L. Quinn is a researcher based in Melbourne, Australia. She completed her Bachelor of Arts / Bachelor of Music combined degree at the University of Melbourne in 2011, with Honours in Psychology completed at Federation University in 2013. Her honours thesis investigated the experience of choking in music performance. She previously worked as a piano teacher, and currently volunteers her time to assist in the running of music workshops at youth-specific mental health facilities in Melbourne, focused on early intervention. She is currently employed by the University of Melbourne in a research capacity, working on two randomised-controlled trials in youth depression.

a.lopatecki@gmail.com

Bernd Redmann was born in Bamberg and studied composition, music theory, music pedagogy and musicology, in Germany and in Austria. In 1997, he was awarded the composition scholarship "Cite des Arts" that took him to Paris, and in 1998, he received his doctorate in musicology from the University of Salzburg. Between 1998 and 2005, he was professor for improvisation and theory at the Cologne University of Music. Since 2005, he is professor at the University of Music and Performing Arts in Munich. Between 2011 and 2014, he was vice president of this institution and now, since 2014, he is its president. Redmann has composed and performed more than 50 pieces, among them works for orchestra, ensemble, chamber music, vocal music, music for media projects and an opera. His publications include a book on music analysis, articles on Beethoven, Schubert, Brahms, Schenker, and Riemann, and essays on improvisation, instrumentation, and the music history of Salzburg.

bernd.redmann@hmtm.de

James Renwick studied clarinet at the Sydney Conservatorium of Music before completing a musicology degree at the University of Sydney. Subsequently he taught woodwind performance, before turning to a research collaboration with Gary McPherson at the University of New South Wales. This resulted in a PhD investigating relations between motivation and self-regulated music practising behaviour. He was Lecturer in Music Education at the University of Sydney, researching diversified models of performance study for preservice music teachers and the motivation of jazz and popular musicians. Recently he shifted his focus to primary-school (elementary) education with the New South Wales Department of Education.

Contact: g.mcpherson@unimelb.edu.au

Richard Allen Schmidt (1941-2015) would like to be remembered for the following things: he was 3rd team all American in rings in 1962; he was an avid, competitive sailor in numerous classes of boats, winning multiple championships; he competed in (and completed) nine marathons and a multitude of other long distance running races; he was Professor Emeritus at UCLA in Psychology, earlier in Kinesiology; he mentored many PhD students during his long academic career; he authored four textbooks, one of which continues to be the premier graduate-level text; he published a high rate of peer reviewed research articles, including the Schema Theory (1975), which earned him a Citation Classic; he was honored as a Distinguished Scholar award by a group of peers and received two honorary doctorates in Belgium and France; in the late 90's, he joined Failure Analysis Associates, and then, in 1998, he started Human Performance Research to offer consulting in human performance as an expert witness.

He passed away in 2015 after a long battle with neurological degenerative ailments, CBD (corticol basal ganglia), PSP (Progressive supranuclear Palsy), and APD (atypical Parkinson's disease). Or, as he would say, he died of TLAs (three letter acronyms).

Simone Spangler is a PhD candidate at the University of Music and Performing Arts Munich, and supported by a Bavarian State Ministry of Education and Cultural Affairs, Sciences and the Arts postgraduate grant. She graduated in music education at the University of Music and Performance Arts Munich, and in sports education at the *Technische Universität München* in Germany. Her current research in music psychology focuses on motivation, wellbeing, practice behaviour and performance. She is also member of the research network "Multizenterstudie," which is conducting a long-term investigation on "Psychological and physical health and preventive behaviour of music students."

info@simonespangler.com

Ingolf Turban commands a large body of masterpieces for violin, some of which have rarely been performed on stage. His virtuosic skill and affinity for Niccolò Paganini's compositions, as heard in his recording of the six violin concertos and the TV documentary "Paganini's Secret" received critical acclaim. He has performed as a soloist in concert halls such as the Berlin and Munich Philharmonic; the Kennedy Center, Washington; Avery Fisher Hall, New York; Tonhalle, Zurich; Goldenen Saal, Vienna; and the Scala, Milan. Turban has worked with conductors Sergiu Celibidache, Charles Dutoit, Lorin Maazel, Zubin Mehta, Yehudi Menuhin, Jun Märkl, and Marcello Viotti, and has documented his extensive repertoire in over 40 albums. He is founder of the Chamber Orchestra "I Virtuosi di Paganini" and professor at the University for Music and Performing Arts in Munich.

ingolf.turban@hmtm.de

Verena Wagner is research associate at the Department of Psychology of the University of Graz and the *Institut für Begleitforschung*, working on different projects within the research field of Work-, Organizational and Environmental Psychology. Since 2012 lecturer at the Ruhr West University (Bottrop), Germany (degree program: human-technology interaction). Verena's research activities include sound perception and sound design, hearing impairment at (elderly) employees/elderly people, Occupational Health Psychology, Human Factors and Human-Machine-Interaction. She studied psychology at the University of Graz (diploma/equiv. master degree and doctoral degree) and instrumental/vocal education at the J.J. Fux Konservatorium, Graz (diploma/teaching certificate) and University of Music and Performing Arts Graz (bachelor degree).

verena.wagner@uni-graz.at

Dr. David Wasley is currently Senior Lecture at Cardiff Metropolitan University. He has been involved in the research, education and provision of physical activity and exercise to enhance health and well-being since completing his BSc (Hons) in Exercise Science at the University of Brighton. He has taught and researched classical musicians' experiences for the last 15 years starting with his PhD that investigated the effects of physical activity and exercise on the psychophysiology responses of musicians as part of the 'Zoning In' project at the Royal College of Music (RCM), London. He is a co-investigator on the 'Musical Impact' project based at the RCM which includes working in multiple UK Conservatoires.
dwasley@cardiffmet.ac.uk

Aaron Williamon is Professor of Performance Science at the Royal College of Music (RCM) and Director of the Centre for Performance Science, a partnership of the RCM and Imperial College London. His research focuses on skilled performance and applied scientific initiatives that inform music learning and teaching. Aaron is a fellow of the Royal Society of Arts and the UK's Higher Education Academy, and in 2008, he was elected an Honorary Member of the RCM.
aaron.williamon@rcm.ac.uk